LIFE IS A BLESSING

CLARA LEJEUNE-GAYMARD

LIFE IS A BLESSING

A Biography of

Jérôme Lejeune

Geneticist · Doctor · Father

Translated by Michael J. Miller

JEROME LEJEUNE FOUNDATION
Paris

THE NATIONAL CATHOLIC BIOETHICS CENTER
Philadelphia

Originally published in France by Éditions Fleurus, Paris, as *La Vie est un bonheur: Jérôme Lejeune, mon père*, by Clara Lejeune-Gaymard, 1997. Translated from the French by Ignatius Press, San Francisco, by permission of Éditions Fleurus, 2000. English translation reprinted with a new foreword by The National Catholic Bioethics Center, Philadelphia, by permission of Éditions Fleurus, 2010.

Cover design by Heather Anne Ermine

To Mama,
faithful queen in my father's heart.
To my brothers and sisters, who shared
the same blessing.

Contents

Foreword

I am honored to introduce you to this biographical memoir written by Clara Lejeune-Gaymard on behalf of her father, Dr. Jérôme Lejeune, on the occasion of the book's re-release in the United States. I had the privilege of knowing Professor Lejeune personally. A pediatrician and geneticist born in 1926, Jérôme Jean Louis Marie Lejeune was truly a giant of the twentieth century, in both his accomplishments and his ability to combine his faith with his scientific investigations.

In 1959, Professor Lejeune discovered that the presence of an extra twenty-first chromosome causes Down syndrome (trisomy 21). He went on to identify other chromosomal disorders, including cri-du-chat syndrome, and did pioneering research on chromosomal anomalies in cancer. But his passion was always the welfare of mentally handicapped children. He died in Paris, France, on Easter Sunday 1994.

Dr. Lejeune was one of the most generous men I have known. He was especially generous here in America as he shared his countless scientific talents with us in defense of life. A number of years ago, he testified in a widely publicized "frozen embryo" case in Knoxville, Tennessee, and it was his testimony that led the trial court judge to base his decision on the fact that life begins at conception. Countless times Dr. Lejeune flew to the United States on short notice to testify in abortion-related court cases. It was my privilege to be with him in an abortion-protest trial in Wichita, Kansas.

Dr. Lejeune was a strong and yet very gentle man. He was also known in America for his ability to recite poetry in English—a second language he picked up by reading a book! He was chosen by Pope John Paul II as the first president of the newly established Pontifical Academy for Life. Unfortunately, Dr. Lejeune's death prevented him from serving the academy. He was an incredible role model to people like me, particularly in his ability to combine his Catholic faith with his outstanding abilities as a scientist.

In this charming story, Clara Lejeune recounts her personal discovery of her father as a wise and loving parent, a compassionate physician, and one of the world's great scientists. Her sense of wonder throughout this awakening is paralleled by Dr. Lejeune's own obvious humility. He was a man without guile. His fame existed solely for the sake of advancing the cause of serving others.

I hope and pray that this book will provide you an insight into this great man who was the leader in Catholic medicine in the twentieth century.

Thomas W. Hilgers, MD
Director
Pope Paul VI Institute for the Study of Human
  Reproduction
Omaha, Nebraska
USA

LIFE IS A BLESSING

Preface

How does one write a book about one's father? His life is at the same time too near and too far for the solemnity of the usual biography. Too near because affection can scarcely maintain a critical point of view; too far because his story is not ours, even though we are, from a certain moment on, intimately involved in it.

And so this is the beginning, quite simply, of the picture book of a child who is remembering, looking through the innocent eyes of that first and unforgettable love.

Happy people have a story, too, but they don't tell it —as though happiness were content with being lived. Malraux used to say that he detested his childhood; I adored mine. I had as my father an extraordinary man who, acting from conviction, chose a cause that was lost from the start, a pessimist whose realism was inspired by a formidable hope.

In this world, where there is talk only of suffering, misery, and injustice, how does one say as well that life can be beautiful, very beautiful?

It is the child speaking. All mothers and all fathers have such a child, who remembers the old days when his parents seemed invulnerable, like infallible guides along the path of life, quite simply because they embodied the love of life and the beauty of love.

The child who sees that his parents love each other

understands everything. He understands that nothing of value is accomplished if it does not bear fruit in the heart of that love. He is there because his parents loved each other one day, one night, thus bringing about the miracle of life. He is happy because his parents still love each other. He is happy because his parents are happy that he is there, a witness to their love.

I'm eight years old; I'm tucked in my bed, near the door, in the room that I share with my sister Karin; and it is night. After the family prayers, Papa has come to our rooms to kiss each one of us and to say goodbye. Tomorrow morning, at dawn, he leaves on a trip. I pray, and I'm afraid. I'm afraid of never seeing him again. The fear of his death rocked my childhood to sleep. It was a loyal fear of a child who trusts but still trembles.

It's funny; as a child I never feared that Mama might die. With her indestructible health and her legendary strong will, the only thing she could be made for was life. Maybe also because she was always with us. She did not leave for countries with strange names to give conferences.

We awaited my father's return by counting the number of beddy-byes. For distant journeys, Mama had invented another method. On a piece of paper hung above the bed there were some circles. Beneath the first one we would draw a line each night. By the fifth day it was a little stick figure, a cartoon man. Papa would return when one, two, or three were drawn. And at school, in front of our astonished classmates, we used to declare peremptorily, "Papa will be back in two little cartoon men."

And Papa always came back, with stories full of people, and landscapes, and encounters, and sights that he saw, to which we listened in amazement and disbelief. Our costume bag preserves a few souvenirs of his travels: a kimono, Austrian skirts, Turkish slippers, silk scarves from China, some painted spoons from Moscow. . . .

But is he really gone forever? I don't believe it. In his own way he keeps on adding to our box of treasures. By his death, first of all, which he experienced as a victorious resurrection on Easter morning, after a long agony that had begun on Wednesday of Holy Week—but I will tell of that later. President of the Pontifical Academy for Life, he died thirty-three days after his appointment was announced. The Pope, who was grieved by his passing, would later say to my sister Anouk, "Humanly speaking, we needed him so much. But maybe this is a gift that he has given us for the Academy and for all this prolife work. Didn't Christ die on the Cross to save us?"

Ever since he has been upstairs there have been more and more signs. The testimonies of his friends, of his patients reveal to us a man whom we didn't know, so discreet he was about his professional life and his vocation as a physician. A friend whom I had not seen for ten years admitted to me, "I didn't know your father very well, but one day our eyes met, and that day we confided in each other. Since his death he has been on my mind, and I think about him all the time. That is all that I came to tell you. I'm here if you need me."

All those who have known him, even if they met him for only a moment, have not forgotten him. In France and abroad, we meet women and men who have crossed paths with him only once, but who remember something he said, a smile, a gesture that touched them profoundly. Signs, there are plenty of them, but Papa didn't like signs.

So we take them as winks that he sends to us in his gentle, humorous way.

How can you say that he is gone when everything reminds us of his presence? It's not that we want to cry—he doesn't leave us enough time for that. We are invaded by letters, telephone calls, trips that Mama makes to replace him, and projects that continue his work with the sick.

On the refrigerator door there is a color photo of Papa, with his big blue eyes. Between two phone calls while sorting the mail, Mama stops and looks at him: "You are too much, you know. Do you see all this work that you're giving us?"

We have to carry on.

Seek and You Shall Find

*And try to render to each person
that fullness of life that we call
freedom of the mind: there is a
task for us, for our successors,
and for their successors.*

My sister is ten years old. At school her teacher says to her, "You know, don't you, that there is a great scientist in your family?"

Embarrassed, Karin remains silent.

"Oh, but there is! Someone very close to you."

Karin rummages around in her memory.

"Maybe my grandfather. He was a veterinarian and invented a treatment for cows."

"You're not even warm, my dear. It's your father."

That is how we discovered one day that he was a well-known, respected scientist. We had heard, of course, about trisomy 21, which gave children a face that they called *mongoloid*. It even seemed that the Mongols returned us the favor and found that *their* children with Down syndrome resembled us.

One day Mama had us wear pretty navy blue coats and put gold barrettes in our flyaway hair. She coached us well so that we would be good. We took the subway, and then we entered a great hall, full of people, with impressive gentlemen in long black robes with wide sleeves that fluttered like exotic birds. Then Papa came in. He was wearing a costume, too. He went up onto the stage and began to speak. That lasted a long time. The people listened. . . . As for us, we passed the time pulling up our white stockings just so. Then Papa kept quiet; there was a moment of silence and then thunderous applause.

We had attended his inaugural lecture. Appointed professor of medicine at the age of thirty-eight, he was the youngest one in France, and it was for him that the first professorship in fundamental genetics in France had been created.

My father had hesitated between two careers: country doctor and surgeon. After earning a baccalaureate brilliantly at the age of fifteen, he launched into the study of medicine, but he repeatedly failed the examination for surgical interns. On the third try he left in the morning to take the test, but, deep in thought, he took the subway in the wrong direction. He arrived late, and they had already closed the room where the examination was being conducted. He returned crestfallen and gave up for good a career as a surgeon.

As an extremely young intern he became the assistant of Professor Turpin and very quickly discovered his real vocation: explaining, understanding, and healing mental handicaps and, more particularly, that strange syndrome that gives such a peculiar face to those who are afflicted by it.

In those days it was believed that syphilis was the cause of mongolism. The disease was disgraceful, and the mothers were presumed to be at fault. Those poor little children, so ugly and retarded, too! The wretched life of their mother is written all over their faces! It is better to cross the street when you meet one; it might be contagious! Certain people, though, think that it is hereditary. But decent folks don't like these mongoloids, and their parents hide them. Anyway, what can be done for them? They are born handicapped, and they will stay that way.

As the assistant of Professor Turpin, my father sees them, examines them with the stethoscope, and tries to understand. It becomes his passion; he knows that there is something to discover, but somewhere else, in a place where no one has looked yet. He reads all that he can find on biochemistry and genetics, and because scientists are publishing more and more often in the English language, he learns English in a few months by the Assimil method. Very soon he is speaking it fluently.

Modern genetics did not exist yet, but it was already known that man has only forty-six chromosomes and that the ape has forty-eight. My father intuited that mongolism is the result of a genetic accident. And it was while using a makeshift microscope, dating from 1921 and repaired with tinfoil, that he made the discovery. He knew that chromosomes function in pairs, one transmitted by the mother, the other by the father, and that there are twenty-three pairs of them. The pairs are identical except for the sex chromosomes. Boys have an XY; girls have two XX. But they are all wrapped up in a ball of yarn, and a Japanese scientist had just discovered the method of untangling them, of lining them up in pairs and of identifying them easily.

My father began to study the karyotype, the genetic ID card of each individual, with Marthe Gauthier, who had learned in the United States the techniques for making slides of chromosomes. Thanks to this method, he determined that all mongoloids have the same genetic characteristic. There are not two chromosomes from the twenty-first pair; there are three of them. From then on this illness would be called *trisomy 21.*

He might have called it *Lejeune's syndrome*, like so many other diseases that bear the name of the one who discovered them. But what was important to him was restoring the dignity of those who are ill and of their families. Trisomy 21 is a genetic accident, it is not contagious, and syphilis is not the cause of it. From now on people would not cross the street any more to avoid contaminating their future offspring when the afflicted child passed with its mother. From now on families would know that if their child was ill, they were not at fault. The term *mongolism* called too much attention to the physical imperfection. *Trisomy 21* would be from now on the name and the true explanation for a disgrace that was all too visible and unacceptable.

This discovery exculpates the parents. Their child, whom they love in spite of everything, carries on his countenance and in his mind the consequences of a genetic error. But it is indeed their child, who resembles them.

Often the mothers use a very revealing phrase to speak about their child who has trisomy: "He isn't finished." It is the finishing touches, the overall completion of the individual that are imperfect, as though a work of art had been fashioned with faulty instruments, as though the music of life had been recorded on a scratched record.

And that is in fact what happens. The chromosomes of someone with trisomy 21 are perfectly normal. But the rendition is bad: the same few bars play over and over, slowing down the construction of the individual and causing imperfections. If it were possible, selectively, to silence this extra chromosome, the individual would be like you and me. At the moment that is impossible, but in the last century tuberculosis was an incurable illness, too.

And also incurable was rabies, which used to drive children mad! Some doctors would suffocate them between two mattresses to prevent them from suffering. Pasteur tried to save them, using all possible means. Those who meant well in sending them off to the kingdom of the dead, did they make scientific progress? No, it was the one who could not accept the thought of giving up when confronted with sickness, suffering, and death. A doctor, a research worker, is an intelligent mind tenaciously clinging to hope. There is always some way of understanding, of going farther in order to relieve pain and to improve life. Death, to be sure, will show up soon enough.

My father made this first discovery of an illness caused by an aberration of the chromosomes in the month of June, a few days before taking his wife and his children to Denmark. At the wheel of the automobile he kept repeating, "I've found it! I'm sure that I've found it!"

After returning to France he confirmed what he had intuited and began traveling from scientific conference to scientific conference. At that time he was already a young geneticist well known for his studies of the effects of nuclear radiation, notably in the corridors of international agencies like the United Nations. He presented his discovery to the greatest American researchers. But he was very young, his scientific field was new, and they would listen to him and smile.

It was only in January of 1959 that he published his discovery in collaboration with Marthe Gauthier and Raymond Turpin. During eight months he explained what he had found to anyone who was willing to listen. Anyone at all could have stolen his discovery from him, but nobody believed him.

He resumed his practice of treating patients, with the help of Marie-Odile Réthoré, today a member of the [French] Academy of Medicine, who would be his assistant in all of his work. As for the administration, genetics still didn't even exist, and so for several years they continued to sign fictitious tests for syphilis so that the laboratory would have permission to function.

"The administration did not have any terminology for karyotypes or for the new genetic tests that we were doing, and so we cheated. It makes you wonder about the reliability of hospital administration statistics. Of course the families knew nothing about it. We, on the other hand, knew very well that syphilis had nothing to do with trisomy, so why plant doubt in their minds, when it was simply a matter of keeping the lab running?" So Marie-Odile Réthoré tells it.

After this landmark publication come other discoveries: the *cri-du-chat* illness, monosomy 9, trisomy 13, and so on. And fame. It becomes obvious that he will receive the Nobel Prize. He has not simply made a fundamental discovery: he has opened the doors of genetics to new generations of researchers.

But Jérôme has become a medical researcher by necessity. Because before he could attempt to heal his patients, it was necessary to understand their illness. Would he have to continue along this pathway to knowledge, which interests him passionately but which brings no short-term relief to the patients? These patients whom he receives, with their parents, three times a week, during his long office hours: he is fighting for them, and he knows that they cannot wait.

So he launched into a less prestigious line of research. One day he returned from his appointments and said to Mama, "I could spend years discovering the genetic causes of many illnesses, and I could keep on studying even rarer diseases. But I am convinced that everything is interrelated. If I find out how to cure trisomy 21, then that would clear the way for curing all the other diseases that have a genetic origin. The patients are waiting for me; I have to find it."

From then on he devoted all his time to the attempt to understand the biochemical mechanisms that result in mental and behavioral disorders.

A precursor—that he was when he claimed that autism is not a psychiatric illness due to the mother's bad conduct, but that it, too, probably has an organic cause. A precursor he was also when he understood the essential role that folic acid plays in the development of little children. He would give it to his patients, who would then improve. He prescribed it for his daughters when they were pregnant. His scientific colleagues sneered, "Your studies of folic acid lead nowhere. You're making a big mistake." It is true that meanwhile he had taken a stance against abortion and that everything he did from then on would be criticized in advance.

Nevertheless, today most gynecologists prescribe folic acid for pregnant women. Ten years after he was able to show clinical proof for it, the scientific press and even the press at large came to the conclusion that folic acid was undeniably effective in preventing spina bifida. No one cited his research, but he didn't care. The important thing was to see his intuition confirmed by others. It meant that he had not worked all those years for nothing.

But a researcher's life involves numerous disappointments as well. How many times he returned home with the sense of having made a major discovery. Sometimes he was mistaken; often it was a matter of a little progress, but nothing conclusive.

My father's research was a combination of genius, puttering around, and imagination. He would often say, "The skill of a researcher when confronted with a difficulty is to keep turning around until he finds the door that opens." But research is like war: everything is in the details.

He also experienced some pitiful defeats. For example, he had put all his hopes for a very important breakthrough in a biochemical substance and was testing it with his patients. Some of them received a placebo, and others the real treatment. Two months later he saw the patients again. Those who had been treated showed a development of the cranial perimeter much greater than that of the others. The parents, too, thought that they were doing much better. But then two months after that, on the contrary, the cranial perimeter had not budged. What a disappointment!

It took him some time to understand why. He usually saw his patients at the hospital, but it so happened that the first time he met with the patients who were being treated, the meetings were held in the faculty offices. Now the faculty's tape measure was shorter by one centimeter than the one at the hospital! What he had hoped for was a mirage. He decided to laugh about it. He wrote this anecdote down himself, one summer evening, without telling us. We discovered it after his death. No doubt it was for him a formidable lesson in humility.

A researcher's life is made up of shadows and light. He drew his strength from the gaze of the sick children, who put all their trust in him. A trust that was staggering, at the heart of their suffering, total and without reservation. He was their doctor, in a certain sense their property. *"Mon professeur"*, as they used to call him, would make their lot a little less unhappy, that was certain.

Faced with such meek submission, what was there to do except to continue unceasingly? He always had the feeling that what he was doing was not enough. He liked to quote the reply of Vincent de Paul when the Queen asked him, "What must one do for one's neighbor?"

"More!"

A Child's Look

Let him live: he will think.
It is the destiny of men.

Papa has something of a mustache, which he trims carefully every Sunday with a little pair of nail scissors. Said scissors are hidden behind the door of a cupboard in the bathroom to prevent a child from using them to cut a piece of wood or a wire. (It appears that the scissors don't like that.) He has his hair combed back, and his large forehead is framed by graying temples. I find that that gives him an intelligent air. But most of all Papa has a look. His big blue eyes, a bit protruding, which sparkle with intelligence and humor, gaze at you with an infinite tenderness. Who said that blue eyes look at you coldly? His are very kindly. Nevertheless, they are demanding, too, because they love truth. They look, untiringly, for the why and the how of what they see. But how can there be so much goodness in the look of a scientist, a researcher?

The answer is simple, and we, his children who see him every day, find it obvious. My father is a man of contemplation and wonder. He often explains to us that the only real difference between a man and an ape is the capacity for wonder. Loving, animals are capable of that: they know about faithfulness, tenderness, and even grief. They can understand, too: they understand language. Doesn't a dog come when someone whistles? They have signs and codes and know how to use certain objects as tools.

But admiring a sunset, contemplating beauty, being aware of the Infinite, and hence being able to reason about the human condition—only man has that grace.

Some time before his death, at the hospital, he, who very rarely watched television, wanted to see a program devoted to the mission in orbit that was supposed to repair the flaws in the Hubble satellite. He was thrilled that so much energy, so much mental ingenuity, and so much money, too, would be devoted to this mission, which had only one goal: knowledge. That for him was the true human genius.

My father was thin and of medium height. His natural elegance was not reflected in the way he dressed; that is the least that one can say. He kept his suits until he wore out the seat of the pants on his bicycle seat and used to drive our mother to despair when, after ten years, he didn't think it was necessary to buy a new suit. He owned a handsome navy blue double-breasted suit that he bought in the early sixties on the occasion of his appointment as professor of medicine. He would wear it until his last appearance at the Academy of Medicine, a few weeks before his death.

Whenever one of his daughters burst out laughing, "Papa, you are so old fashioned!" he used to reply, "Oh, no, my dear, I am a precursor; this will be the fashion in ten years!"

Everyone laughed and made gentle fun of him.

How many times the sympathetic fathers and mothers of our girlfriends said to us, "It certainly must be difficult to be the children of a famous father, with such a strong personality, with the conferences that he gives all over the world . . ."

They didn't know.

They did not know that this great professor of medicine with such fixed ideas was above all a father and a husband. Being around him and Mama, we had a fabulous childhood.

All the while that my father was spending his days with his patients and their parents, easing their sufferings, while he spent hours with a microscope and a computer looking for the key to the cure, while he crisscrossed the world in order to share his findings and his brilliance with an uncalculated generosity, while he wiped off the spittle of contemptuous know-it-alls, while he meditated on human nature and the divine plan, he was loving us, too.

My father came home every day to eat the midday meal with his children. For our sake he gave up all those "business luncheons" that enable one to maintain contacts that could be useful in the future career of a researcher or a promising physician. In the evening he was back at seven-thirty for dinner. Few children today have the privilege of sitting down to three meals a day as a family with both of their parents. At noon the meal was quick, since we all had to get back to school or to work. In the evening we had more time, and Mama would always start off the meal by asking Papa, "Whom did you see, and what did they say?"

Papa would then tell us about his day, his encounters,

his work. That often led to discussions in which we would reconstruct the world. Once we had reached adolescence it really widened our horizons, and my parents appreciated very much our arguments about ideas.

Quite often my parents had guests over for dinner or supper: some friends, colleagues of Papa, or people who came from all over the world.

From an early age we were included at these meals, when we wanted to be. This enabled us to become acquainted with and to appreciate a great number of personages, both French and foreign, who remember today the old house dating back to the Middle Ages: a house of our own, though not much in the way of comfort, where the children reigned, together with a charming disarray and a hospitality without limits.

It was the house of our dear God, where all friends were welcome to eat, to sleep for a night or for several months. Mama was completely lost when, by way of exception, she had to cook for only two or three people. Every year there were dozens of Danes, from the village where Mama was born, who came as tourists to France and were welcomed, for the duration of their stay, at the *pension complète rue Galande* (at the Lejeune bed and breakfast on Galande Street). Papa cut a fine figure in the midst of these giants, who ate three baguettes a day and could pronounce at most three words in French, and who were extremely well mannered when they did.

Several of them, I'm sure, found the living good, the welcome warm, and the company pleasant. They could stay all winter just waiting to be chased out by other visitors. The worst thing, no doubt, was when they would try to compensate Mama, who didn't want to hear of it. So then they would conspire with her to repaint the living room or the kitchen. Papa would return in the evening and see the apartment turned upside down. He, at least, knew very well that the crack in the wall over the sofa would reappear in a month, since it was a very ancient house with walls that warped. But with his kindness and his incredible patience he would say nothing.

I remember also coming back from weekends, vacations, or various expeditions, when we ended up at the house without warning, ten or fifteen of us, all dirty, to have dinner on Sunday evening. I can still see Mama, greeting us with open arms and bringing food out of the refrigerator to meet the invasion. And Papa would come down the stairs with his big smile to keep us company and to hear the story of our adventures.

Then, when the meal was ready, he would disappear with my mother so that we would be free to discuss things at our leisure among friends. Except, and this happened more frequently, when our friends held him back so as to hear him speak, captivated by his learning, his eloquence, and his very wide knowledge of people and of things.

Like all papas in the world, he used to call us by silly, endearing nicknames that sometimes embarrassed us. He put all his paternal tenderness into it, but also his ability to tease. He would say to us, "You see, it keeps me from mixing up your names."

Never, really never, did Papa refuse to answer one of our questions because he didn't have the time. I often think —I, who sometimes beg my children to leave me in peace

—of how thin we must have worn his patience and the
soles of his shoes when we were little. I can still see him
concentrating intently on what he was reading or writing. We would come in as charming as can be: "Papa,
can you fix my cowboy outfit?" When he saw that it was
very important to us, he did not say, "Later on; you can
see that you're disturbing me." He left the manuscript of
his lecture or his scientific calculations to repair a bicycle tire, string a bow, glue a broken doll, and answer the
most incongruous questions. "Papa, did you fight in the
Hundred Years War? . . . Why are people born? . . . Why
does it rain? . . . What are the stars for?" We always got
an answer, and so for that matter did *our* children when
they took up the refrain with their grandfather. And we
are very unhappy now that we can no longer say, "We'll
go ask Papa."

Yes, it's true that Papa knew all these things and many
others, too. He had what was once called *la culture de
l'honnête homme* (the education of a respectable man). He
knew how to read Greek and Latin, was acquainted with
all the classics, had an appreciation for painting and music, and nourished his mind with philosophy and theo-
logy. It was impossible to stump him in history, and he
had a passion for antiquity.

He liked recreational mathematics and kept up a regular
dialogue, over several decades during his spare time, with
a priest-mathematician by the name of Masceroni (whom
we stupidly called Macaroni), who had lived during the
eighteenth century and had invented compass geometry.
The purpose of the exercise was to construct geometric
figures using only a compass with no ruler, thus without straight lines. We saw him one evening triumphant,
because he had succeeded in constructing with the com-

pass a complex geometric figure, whereas Masceroni, the expert on the subject, concluded in his book that it was impossible. There was a side to him that said, "It is that much more beautiful because it is useless", which furthermore had a way of irritating Mama.

My father loved to help others understand. He considered knowledge not as a sign of power but as a communion. Then too, he did not explain; he recounted. And what child doesn't love to have stories told to him! He had a talent for parables, and the book of life that he read to us was made up of colors, images, and sounds that suddenly became animated. It is too bad that our science and math teachers did not always appreciate the picturesque descriptions that we derived from these paternal explanations. They occasionally cost us an "F". Nevertheless, we were happy. Because without learning the lesson we had understood. And then we could tell Papa that he had got an "F" in math!

But above all, he was a poet. He studied the science of humanity and examined the mirror of the soul through the prism of an artist. He had been surrounded by a mother who was a musician and two brothers who were painters, and art was for him the chief expression of human creativity.

His lectures, even in very technical fields, always had their share of metaphor, so as to facilitate understanding and to touch the very heart of the mind. With his pleasant voice he was a formidable orator whose style was almost celebratory, since his poetic talent was combined with subtle humor. The most complicated of scientific theories could then be read like an open book, even by the uninitiated.

Physician of Men, Physician of Souls

*Our intelligence is not just an
abstract machine; it is also incarnate,
and the heart is as important as the
faculty of reason, or more precisely,
reason is nothing without the heart.*

Papa was a very young physician. To make a living, he
substituted during the summer for country doctors. He
left, with Mama and Anouk, my older sister, who was
just a baby, for the region of Gers. He would make the
rounds of the villages. In later years his memories of those
days would always remain vivid, brimming with memories. There he learned about the work of a general prac-
titioner but also about life and the heart of the people.
He would always be a little nostalgic about this work as a
country doctor, which was such a rich experience, both
humanly and professionally. It was the work that, deep
down, he would have liked to do permanently.

One story left a profound impression on him. The roads
in that region are difficult, narrow, and meandering. He is
called to help deliver a child. He arrives at a large farm, and
the family is around the bed of the young woman, who
has gone into labor. Everything seems to be going well.
The future mama is a woman in good health, strong and

brave. But he is struck by the somber and sorrowful expression of her mother, who is assisting him in the preparations for the delivery. The birth takes place smoothly; it is a strapping baby boy. But there is a twin who has already died. Papa does not want to perform the curettage at the house; the mother would suffer too much. He decides to take the young woman and her husband to the hospital.

Before saying goodbye to them, the old woman follows them and, outdoors, says to him, "She is going to die." My father protests. She answers, "I know it. She is going to die, no matter what, and you will not be able to do anything about it." He leaves, and an irrational anguish pursues him while he is driving the young couple to the nearest hospital.

Once they arrive, the young woman balks. She wants to go to a clinic that she knows, further on. She stubbornly insists, and Papa again takes the wheel of the car. The road is very bad, and with his old 4CV it seems to take him forever. At last they arrive and have the woman admitted, and the medical team on duty takes over. The following morning, when he returns, she is dead. Her mother weeps, saying, "I knew it; I saw it." He, the doctor, who had performed many deliveries, which were often much more difficult, had had no foreboding at all.

A physician: that's what my father was, to the depths of his soul. Everything about this vocation was attractive to him. Helping sick people, dialogue with families, but also the scientific aspect. Understanding the human body, its subtle mechanisms, the origin of life. For him it was an object of study, but also of unending wonder. What a marvelously ingenious and complex machine is this body that makes us live!

Physician of men, of those who are mentally disadvantaged, he lived at the heart of an unspeakable suffering, which prevents the sick person from being fully himself. Very quickly, and without realizing it, he became a physician of souls.

Often he was called to the side of young parents who had just learned that their child was not like the others. Many times he was overwhelmed to see that the drama of a handicapped childhood is often heightened by the cold, technical, and often cruel manner in which the diagnosis is announced.

"Your child is a monster; he would be better off not living."

"You have seen the head; the child is mongoloid. I advise you not to keep him."

Then there is the feigned ignorance: "On the contrary, madam, your son is quite well; you have no reason to worry."

Then the fateful morning when the head physician comes, accompanied by a host of interns. They bend down over the child with a serious mien, use scholarly terms, and then leave again: "We are going to run a series of tests." Several days of anguish, only to learn sorrowfully from an anonymous laboratory a truth that is so cruel to discover.

He heard thousands of such stories. Other parents, fortunately, had better support when the handicap was announced. But there was no end to his visits, which lasted for hours, with these distraught young parents, in order to talk with them, to tell them the truth, the whole truth, about their child's condition. What would his future be? What would he be able to do? And to counsel the parents on what they could or wanted to do for their child.

After the death of Jérôme Lejeune, the testimonials

came pouring in. Having a great respect for professional confidentiality, he rarely spoke of his meetings with his patients. At the very most, on certain days when he came home anxious, we would ask him, and he would say to us, "I have a patient who is not doing well. I'm worried."

"It is truly legendary, the relations that Professor Lejeune had with his patients", Professor Lucien Israël would later say.

Being the head of his department, he could have had a private practice on the side like all of his colleagues, whether at an office downtown or at the hospital. He always refused to do that; as he saw it, he was at the service of the patients, and the state, by paying him a salary, enabled him to live decently without seeking other ways of making money. The patients, who came from all over the world, were sometimes very surprised that they had to pay only 130 francs, the cost of consulting a specialist. We have never been rich, but we were not poor, either. The important thing for him was being able to provide us with a decent education. But as a family we never went skiing, for example. Our means did not permit it.

One day my husband and I were invited to the home of a renowned professor of medicine, now retired after practicing at Necker Hospital. He was a lover of antiques, and he had made himself a fortune thanks to his reputation as a leading specialist. He smiled as he greeted me and said, "I have a great deal of respect for your father, but the difference between him and me is that I used to arrive at the hospital in a Ferrari, while he came by bicycle!"

When a couple would come to see him for the first time with their child, often they had just learned about the handicap. The parents were distraught, anxious, exhausted. We have heard the story countless times:

> The future for us was dark. We felt at the same time unable to keep this child, who was going to destroy our happiness, and unable to abandon it. We went through all the tests, and finally they gave us the diagnosis. We ended up detesting this child and detesting ourselves because of it. After all, it wasn't the poor child's fault. Then we went off to see this famous professor in a big hospital in Paris. It was both intimidating and reassuring. At the same time we thought to ourselves that it was no use. After all, the child's life was ruined.
>
> The professor greeted us with a smile. He was courteous, friendly, but respectful. He turned to the baby, asked his name, and said to him, "Little Pierre, will you come with me?" He took him in his arms, asked the mother to put on a hospital gown, and offered her a seat. She sat down; he put little Pierre in her arms, sat down across from her and the father, and with a stethoscope examined the child on his mother's lap. For us these simple gestures were like a revelation. It wasn't a patient this doctor was examining; it was our child.
>
> Then he explained everything. What this illness is, what the future will be for the child and for us. He reassured us, responded to all our questions and fears.
>
> Before leaving us he said to us, "If you wish, for your next appointment bring his older sisters along. They, too, have the right to know and to understand." We left with our baby, all of us much calmer. He helped us to discover our love as parents.

Thus he was the physician in attendance when entire families came to the consultation of the little brother or the little sister. He remembered perfectly each one of his patients and called them by their first names. He followed their development, their little troubles, and their victories.

I remember. Often, when we were children, his blue eyes became still, and his gaze would rest upon us for a long time, with tenderness, with gratitude. We loved his silent love. Now I know what he was thinking, "How fortunate my children are, not to be sick. Thank you, Lord, for granting me this blessing."

He savored this infinite joy that he, more than anyone, could appreciate. We, too, had our health problems, some of them serious. We must have caused him a lot of anxiety at times, but he knew that a handicap is a cross that is carried every day, unceasingly. On good days as on bad days, without respite.

He called his patients "the disinherited". Disinherited, because their genetic heritage was not perfect. Disinherited, because they were the unloved members of this competitive, glamorous society. To see how much he loved them, one might wonder how human beings who were so uncouth looking, some of them, and so little endowed with intelligence could inspire such tenderness. As for the answer, it was Cecilia, who has trisomy, who gave it to us. She had written a little poem in honor of her professor:

> My God, if you please,
> Watch over "my friend".
> My family doesn't like me much,
> But he thinks I'm kind of pretty,
> Because he knows what my heart is made of.
> Oh, sure there are beautiful children,
> But are they really
> When they make fun so shamelessly?

People called him often, at any hour of the day or night, to consult with him, for counseling. The pregnant woman who learned from the sonogram that there was "a risk". Parents whose normal child had to undergo surgery, a difficult treatment, or who were upset about a diagnosis. Quite often friends of friends would call us: "I have a girl-friend who is upset. Could you ask your father what he thinks?" And then they would explain the problem over the telephone. We always answered, "Tell your friend to call him this evening, at such and such a number."

"But she would never dare to disturb him."

"Then give me her number, and he will call her this evening."

People were often surprised at how available Papa was. But he had taught us one thing: "When the parents are upset, face to face with a sick child, we do not have the right to make them wait, not even one night, if it is possible to do otherwise." We had some Christmas celebrations, Saturdays, and Sundays interrupted by these phone calls or visits, but it was such a small thing in the face of suffering.

His advice or his counseling always prevailed, no doubt because he never concealed any of the truth. I remember a woman, pregnant at age thirty-five with her first child after having a great many difficulties, who, in the fourth month of pregnancy, came down with a serious case of chicken pox. She panicked and felt that she was incapable of accepting the birth of a handicapped child. She consulted fourteen different doctors, who all told her, "It would be better to have an abortion, madame." But this child—she had waited so long and wanted it so much. She did not see how she would be able to give birth to another.

Despairing about the whole situation, she used several intermediaries to contact Papa. That very evening he called the woman. They had a long telephone conversation. The decision was made; she kept the baby. Today she is a delightful little girl.

We have had similar testimonies, numbering in the hundreds. What is surprising is the impact of the convictions and counseling of a man who, quite simply, would let the parents make a decision once they had been informed. Before, they were hesitant, plagued by uncertainties; they would listen to all the opinions, often contradictory ones. He did not choose for them; he showed them their responsibility as parents but gave them also all that they needed to choose freely.

One Plus One Equals One

*To dissociate the child from
love is, for our species,
a methodological error:
contraception, which is to make
love without making a child;
artificial (in vitro) fertilization,
which is to make a
child without making love;
abortion, which is
to unmake the child;
and pornography,
which is to unmake love:
all these, to varying degrees, are
incompatible with natural law.*

There is an elementary operation, the mathematical precision of which varies according to the circumstances and the environment; its coefficient is sometimes very important.

In our house, in a very prosaic manner, one plus one makes one, then one plus one makes seven, and now one plus one makes thirty-five, and as yet not all the children have had their final say! What a strange law it is that allows two spouses who unite in one flesh to multiply their love in the flesh. Our parents had five children, the twenty-fourth grandchild saw the light of day before summer started, and another is on the way.

My father often said while smiling that, in his youth, he dreamed of marrying a tall blonde whose name would be Dominique. He was captivated by a little woman with big dark brown eyes, the air of an Eskimo from the great north, and long hair so black that it had highlights of navy blue. She had a strong accent from the Baltic Sea and learned French on the wooden benches of the *Bibliothèque Sainte Geneviève*. As the family legend would have it, this library was the place of their first encounter. Mama needed a pen, and Papa gave her his. Years later they still wrote love letters to each other with the pen from their meeting. When someone asked him, "What is love at first sight?" my father told us with great modesty, "I do not know why the knees tremble and the throat gets dry when one is in love, but, no doubt, what your mother and I experienced is called love at first sight."

They would have to brave the studied silences of a bourgeois family that was uneasy about this darling of a poor young woman, a foreigner, and Protestant besides, who certainly had not had a suitable education. We, their children, look with emotion at the photographs of their marriage in the Catholic church of Odense, the city of Andersen, in Denmark. Mama is very beautiful, dressed in a black suit, as was the fashion at the time, and Papa is radiant, but unquestionably he is more handsome today! It doesn't surprise us that only our uncle and his wife had come by motorcycle for the ceremony.

What's more, they married on the first of May. Mamie (our grandmother) was not superstitious, but all the same said: "That child was born on a Friday the thirteenth and is getting married on May Day ["unlucky" days]. . . . I only hope that nothing happens to her!"

But those are very small obstacles for a love so great. With her intelligence, the memory of an elephant, her intellectual curiosity, and her passion for news, Mama could have had her own career. Papa always said that she would have made an excellent journalist. She chose to live in the shadow of her husband and to raise her children. Deep down inside, she must have known very well that Papa could not live without her and, above all, would never have accomplished all that he did if he had not had her daily at his side.

Then too, there is no doubt that she was always conscious of living at the side of an extraordinary man. But living with an exceptional being demands quite a lot of courage. The affronts, the humiliations, the financial constraints, the friends who abandon you because of your husband's activism. Mama faced them all in her pugnacious and optimistic way, always seeking to defend and to protect the one whom she loved. But when it was a matter of going to the front, she was always side by side with her pro-life soldier. No doubt it was thanks to her, too, that he never laid down his arms, thanks to her determination and her strength of character that Papa often ended his lectures by saying, "We will never abandon them."

Mama was an only child. She was raised alone by her mother, who was widowed very young. All her life she has never stopped recreating a large family. It would be around her and Papa that my father's family would regroup, after having been divided by bereavements, separations, and conflicts. A hurting family that rediscovered the melody of happiness. And every year, from then on, our cousins would join us for a Scandinavian Christmas.

Mama is a woman who gets involved. She loves life, and it rewards her in kind. No mere observer, she experiences political or social events with conviction and passion. She reads all about current events and always keeps up with the latest developments in this affair or another. Her prodigious memory enables her to compare the statements made by a particular politician after an interval of several years. At times this comparison is rather cruel for the person in question. Fortunately for him, it is kept within the bosom of the family.

My parents cooperated with one another completely, and so their rare disputes became the subject of family traditions that would recur periodically. I remember certain conflicts that we witnessed as amused spectators.

We had rhubarb plants growing in our garden. They sprouted along the wall beneath Papa's office, and we used to hide under the large, somewhat rough leaves, which gave off a bittersweet fragrance. Mama used to gather the stalks in the spring and make delicious tarts for us. Papa would protest, "You can see very well that they are not mature; you mustn't gather them until the month of August, or else they won't grow back well next year." And Mama would only shake her head and do it her way. In August she will be in Denmark, and the rhubarb will mature without her, and anyway this has been going on for years now, and the rhubarb still grows, doesn't it? When the rhubarb tart arrived at the table we would try to conceal our giggles while waiting for the familiar scenario, which always ended with an "as you wish, my dear."

In the morning Mama would carefully place in front of the door the trash to be taken downstairs. The first one to go out has to see it and will be so kind as to take it when he leaves. Papa, lost in thought, would push the voluminous bag to one side and then leave in peace.

"Didn't you see the bag that was to be taken downstairs?"

"What bag? But Sweetheart, you should have told me, and I would gladly have done it."

Having an absentminded husband demands a lot of patience at times.

At the start of the marriage Papa remembered anniversaries and holidays. One Christmas he brought home a present, which Mama opened eagerly. It was a figure made of chocolate. Seeing that Mama was a bit disappointed by such a modest present, he said to her, "Well, look and see whether it has a heart of gold." He had carefully concealed inside it a gold watch.

But more often he would forget the calendar date, the day of the week, and the hour of the day. Mama used to warn him a week in advance, "You won't forget my birthday." Wasted effort. So, when the day came, Mama would produce a pretty package and tell him, "Here, my dear, is what you're giving me."

One day at lunch Papa arrived solemnly with a mirror and in front of us all, to our astonishment, looked at his reflection, waited for silence, then said, "Happy birthday, Jérôme."

Of course, we had all forgotten his birthday, but really he was exaggerating!

I was eighteen years old, I had just received my driver's license, and Mama sent me to the dealer to buy a new car. Check in hand, I explained to the nonplussed salesman and left at the wheel of a beautiful, brand-new 405 station wagon.

Every ten years Mama would soften the blow for Papa, knowing that he would protest loud and long that the old car can still run for many years and that it is foolish to spend money thoughtlessly.

We were glad that Mama made the expenditures, because if we had listened to Papa, the seven of us would have lived in a two-room apartment heated with wood, and we would have eaten stale bread, old fruit, and over-ripe Camembert so as not to waste anything.

Now don't go thinking that Papa was stingy. He had absolutely no idea of how much money he had in his bank account. His entire fortune consisted of the hundred-franc bill that Mama gave him from time to time to buy cigarettes or a pack of chewing gum. He was quite simply not at all, and I mean not at all, attached to the goods of this world. He got by with very little, and he was happy. Mama, given her practical side, thinks that modern comforts make life in a large family easier, and so she benefits by them. And we do, too.

"The Story of Tom Thumb"

While scientific progress is discovering
each day a new secret about life,
some would have us believe that we
know with less and less certainty
what a member of our species is.
. . . The legislators of today owe it
to themselves to compose a
declaration that will define this
epoch. Confronted with the tyranny
of democracy, they have to proclaim
the rights of human beings.

In 1972 an initial draft law, called *"proposition Peyret"*, started off the debates about abortion. Under a law dating back to 1920, persons who performed abortions could be sentenced and punished.

This first draft law was concerned exclusively with infants with handicaps that were detected before birth. Why allow individuals to live who will be unhappy and who will make their families unhappy?

Fate sometimes takes turns that are painfully ironic. Two men each made a fundamental discovery that, they hoped, would advance the state of medicine with a view to curing disease.

It was Professor Liley, originally from New Zealand, who first invented the technique of prenatal diagnosis. He hoped that in this way one could detect and treat sick infants at an earlier stage of their development. The other was my father, who discovered the cause of trisomy 21 and was using every possible means to find out how to cure this condition. He, too, was convinced that it ought to be treated very early, in utero.

The two men knew and respected each other. They would become helpless witnesses of the reversal of their respective discoveries. Thanks to amniocentesis and karyotyping (a method for determining the chromosomal characteristics of a cell), the technology was in place for eliminating "undesirable specimens" before birth. Their discoveries were diverted from their original objective.

"*Un dossier de l'écran*" ["Onscreen dossier"], a very popular program in those days, alluded for the first time in the course of a televised debate to the question of aborting pre-born infants who were found to be handicapped. The only ones who could really be recognized at the time were those with trisomy.

The parents of children with trisomy experienced this as a veritable search-and-destroy mission. "What has he ever done, my little boy, that they want to do away with those who are like he is?"

One morning a ten-year-old boy with trisomy came for a consultation. He was crying inconsolably. The mother explained, "He watched the debate with us last night."

The child threw his arms around my father's neck and said to him, "They want to kill us. You've got to defend us. We're just too weak, and we don't know how."

From that day on, Papa would untiringly come to the defense of the pre-born child.

He knew very well how much he would lose in the battle. One of our friends said to us recently, "There are some fights that you just have to fight. You don't always win them."

And Papa, better than anyone, knew where that fight would lead.

A while before that he had gone to an international conference on health in New York. Many years previously he had been named the French expert on "the consequences of atomic radiation for human beings and their descendants". One day, in the inner sanctum of that august UN body, the debate on aborting the unborn unfolded with the usual arguments: the mortality rate from clandestine abortions, preventing deformed infants from coming into the world, sparing women moral and psychological suffering, and so on. Alone in his camp, Jérôme Lejeune took the podium and spoke about the unique child, the likes of whom would never again exist, whose life was being jeopardized by the proceedings going on at that moment. Is life a fact or a desire? He affirmed, "Here we see an institute of health that is turning itself into an institute of death."

He spoke in English and played deliberately on the words "institute of health" and "institute of death". That evening, as every evening, he wrote to Mama and confided to her, "This afternoon I lost my Nobel Prize."

It must be admitted that he had been formed in the school of hard knocks in May 1968. For months he was the only professor who had not missed a single hour of class and who had never been booed or shouted down. His strategy? Listen, never get angry, but do not yield an inch of terrain.

One day when he was working in his laboratory at the Faculty of Medicine, some students who were on strike came in, intending to store some medications that they had stolen from the hospital pharmacy. They thought that the lab would be deserted and were surprised to find him there. Papa suggested a room where they might store the medications, and the students left, happy to have met a professor who was so understanding.

They came back the next day looking for the medications, for "their comrades who were fighting against the National Guard storm troopers".

They were many, and they demanded to have their spoils of war immediately. Papa politely opened the door for them and said, "I would be glad to, but do you have the authorization form?"

He had remembered a nun who, whenever someone asked her an embarrassing question, always answered, "I will have to ask the Mother Superior."

He used the same method. He answered every argument or threat in the same way: "Nothing would please me more, but I need to have the voucher."

There were many of them, and they were very agitated. They could easily have socked him in the jaw and taken the medications. However, weary of the conflict, they left, and at the end of the strike the medications were returned, intact, to the hospital pharmacy.

May 1968 taught him how to defend himself when he was without support. He knew that opponents always respect those who have courage. And he had courage enough and to spare.

With a team of professors—three of them out of the entire faculty—he prepared the final examinations for that year. In July they negotiated with the Minister of Education, Edgard Faure, and, contrary to all expectations, obtained permission to hold the exams in September. By working night and day they succeeded in preparing the examination questions so as to present the minister with a fait accompli. The medical students would be able to take their exams and thus not lose a year of study. Things turned out the same way for all the other students, too.

One of the consequences of May 1968 left him very bitter. The sluice gates of the medical school admissions policy were opened, and he understood, twenty years in advance, the problems that we are experiencing today: "They're training too many physicians. In a few years some of them will be earning the minimum wage or else unemployed. Payment by procedure will be blamed for generating out-of-control medical expenditures, whereas the excessive number of doctors will be the real cause."

No one listened to him in that arena, either. But that wasn't his battle. He was there to defend the smallest and the weakest.

He often said that "a society that kills its children has lost its soul and its hope", and he continued:

For the greatest lesson taught by biology is precisely this debut of the human being in a condition so lowly as to be astonishing, and this manner of fashioning oneself in the shelter of one's mother, hammered out by the indefatigable hope of the throbbing aorta and quickened by the rapid beating of one's own heart. It is the very lesson of unwearied hope. And it is this same heart, which was beating within you on the twenty-first day of your existence, which you must consult to determine your course of action. Every day you will have to achieve again this impossible synthesis between true values and a hard reality. Every day we will have to struggle, we will have to convince ourselves and others, too, and it will be difficult, uncertain, impossible. Just don't forget that last year it was already hard; it was difficult; it was uncertain; it was impossible then.

This reflection alone determines our only possible course of action, which can be summed up in a single phrase: come what may, and whatever may happen to us, we will never abandon the little ones.

Oh, of course he went about it in dead earnest. He pilloried "the chance intellectuals . . . who pretend to believe that it is possible to hold the most contradictory opinions as long as they phrase them elegantly. To excuse murder they have invented the marvelous hypothesis that no one was being killed. They have succeeded in circulating among the general population the astonishing proposition that a tiny, two-month-old man, a little man at the age of ten weeks is neither human nor alive."

He also blamed "the pedantic utilitarians" and ridiculed those who pretended that the embryo is not a human being:

> [As the human mind evolved, it] brought about the appearance of a new language that would finally make it possible to understand something that the human race had been doing for thousands of years without ever realizing it, terminology that would enable it to know how it ever managed to reproduce itself. And it [the intellect] discovered that all one had to do was to think forcefully enough that this thing in its developing phase was human for it to become so. But if one did not think forcefully enough about it, or if someone thought the contrary, the thing would perceive it by a strange sort of perception, by the infusion of some kind of sociologico-parental spirit forbidding it to become human, and the thing then would not become human.

What he was battling against was the refusal to look reality in the face.

"A man is a man is a man." Say that this little man bothers you and that you prefer to kill him, but say the truth. What you are dealing with is a little man. It is not a mass of tissue; it is not a little chimpanzee; it is not a potential person.

And he would tell the true story of Tom Thumb, this little man that we all once were in our mother's womb. In those days, when he used to say this, he was challenged by many of his colleagues in the sciences. Today, the marvels of technology allow each one of us to verify, by sonogram, that everything he said then was scientifically accurate. Thus we now have films of surpassing beauty to show us the beginning of life. But the same people who note with amusement that certain babies grab the amniocentesis needle in their hand and that it is quite difficult to make them let go in the next breath will recommend in vitro fertilization to the parents. The argument that Jérôme Lejeune untiringly defended, "Don't kill it, because it is a human being", is thus rendered null and void. Our schizophrenic society proudly presents at the family dinner table the latest sonogram of the child who is a few months old and yet uses abortion as a simple remedy when contraception fails.

Nevertheless, this is the reason that he was and still is hated so much by certain people. How can one dispute a scientific truth? Certain people have been so blinded with fury at his remarks that they did not realize that Jérôme Lejeune, in setting forth his proofs, was thereby as well, one of the first to reveal to the eyes of the general public the magic of life.

The following text was written in 1973; it sums up

forcefully all of the conviction, scientific certitude, and rhetorical talent that made Jérôme Lejeune an extraordinary defender of life:

> Modern genetics can be summed up in an elementary creed as follows: in the beginning is a message, and the message is in life, and the message is life. A veritable paraphrase of the first sentence of a very old book that you know well, this creed is still the creed of even the most materialistic geneticist. Why? Because we know with certainty that all of the information that will define the individual, that will dictate not only his development but also his subsequent conduct, we know that all of these characteristics are inscribed in the first cell. And we know this with a certitude beyond any reasonable doubt, because if this information were not entirely encapsuled therein, it would never arrive, for no information enters into an egg after its fertilization. . . .
>
> But, one will say, at the very beginning, two or three days after fertilization, nothing exists yet but a tiny mass of cells. In fact, it's only one cell to begin with, the one that results from the union of the ovum and the sperm. To be sure, the cells multiply actively, but that little mulberry that will nestle in the wall of the uterus, is it really different from its mother already? I should think so: it already has its own individuality, and, almost incredibly, it is already capable of controlling the maternal organism.
>
> This minuscule embryo, on the sixth or seventh day, while just one and a half millimeters in size, immediately takes charge of the biological operations. He and he alone stops the periods of his mother by producing a new substance that obliges the corpus luteum of the ovary to function.
>
> Tiny as he is, he is the one who, by a chemical command, forces his mother to offer him her protection. Already he is having his way with her, and God knows that he will not give this up in the years to come!

Fifteen days after the period is missed, that is to say, at the actual age of one month, since fertilization took place fifteen days before that, the human being measures four and a half millimeters. His miniscule heart has been beating for a week already; his arms, his legs, his head, and his brain are already roughly formed.

"At sixty days, that is, at the age of two months, or one and a half months after menstruation stops, he measures some three centimeters from the head to the tip of his buttocks. Folded up, he could fit into a nutshell. Inside a closed fist he would be invisible, and that closed fist could crush him inadvertently without anyone noticing it. But open your hand; he is practically finished: hands, feet, head, organs, brain, everything is in place and will do no more than grow. Look at him more closely, and you will already be able to read the lines on his hand and tell his fortune. Look even more closely, with an ordinary microscope, and you will make out his fingerprints. Everything needed for a national identification card is there right now. . . .

The incredible Tom Thumb, the man smaller than my thumb, really exists: not the one in the fairy tale, but the one that each of us once was.

But the brain, someone will say, will not be completely developed until around the fifth or sixth month. No, you're wrong; it still won't reach its final form until birth, its innumerable connections will not be established until the age of six or seven years, and the totality of its chemical and electric mechanisms will not be running smoothly until the age of fourteen or fifteen years!

But does the nervous system of our Tom Thumb function already at two months? But of course: if his upper lip is brushed with a hair, he moves his arm, his body, and his head as though to escape. . . .

At four months he fidgets so vigorously that his mother perceives his movements. Thanks to the quasi-weightlessness of his space capsule, he makes a lot of somersaults, a stunt that will take him years to perform again in the atmosphere.

At five months he grasps firmly the tiny stick that is placed in his hand, and he begins to suck his thumb while waiting for delivery. . . .

Then why the discussions? Why should we wonder whether these little human beings really exist? Why rationalize and, as a famous bacteriologist has done, pretend to believe that the nervous system does not exist before the age of five months?! Every day, science reveals to us a little more about the marvels of this hidden life, the world of these minuscule people, a world teeming with life and even more charming than the tales told in the nursery. For the make-believe tales were based on this true story; and if childhood has always been enchanted by the adventures of Tom Thumb, it is because all of us, whether children or adults, once were like Tom Thumb in the womb of a mother.

Today it may seem incredible, but this sort of talk was forbidden in the sixties. It laid a guilt trip on the mother; it was unlawful; it represented "the dictatorship of the moral order". Because he wanted to affirm, loud and clear, a scientific truth from which a duty followed, Jérôme Lejeune became involved in a much bigger battle, which was unusually violent.

I was twelve or thirteen years old at the time. On our way to school, which we used to travel by bicycle, my sister and I would pass the walls of the medical school, on which were painted in black letters the phrases:

"Tremble, Lejeune! The MLAC [a revolutionary student movement] is watching."

"Lejeune is an assassin. Kill Lejeune!"

Or else:

"Lejeune and his little monsters must die."

Believe me, that brought our childhood very quickly to an end. These are things that cannot be forgotten even if, during adolescence, there is a sort of play-acting in these street wars, the seriousness of which is not realized.

And it wasn't just words. At every meeting he was harassed, often physically. One time, during a debate at the *Mutualité*, it was impossible for him to take the floor. The audience was yelling, and he was hit in the face with raw calves' liver and tomatoes. He did not step down, but waited for a lull and then yelled louder than the others, "Those who are with me, leave the room!"

After a few moments of bewilderment, the auditorium emptied. There remained about fifteen persons, led by a vociferous Dominican priest, who had placed themselves in a diagonal line across the auditorium to give the impression that they were very numerous. When it became evident that they could now be counted, they left; the people returned to the auditorium, and they were able to begin the conference.

Twenty years later, I have had various opportunities to gauge how tenacious this hatred can be. He who hated no one, who always said, "I am not fighting people; I am fighting false ideas", is, even today, the object of unconcealed fury on the part of those who set themselves up as the apostles of tolerance. Maybe it was that calm strength that walked straight ahead, without worrying about what was "politically correct"; maybe, too, it was that oratorical gift employed to defend the life of pre-born children; maybe it was above all because he had been brought to the pinnacle of science and fame by people who resented him from then on because he made their political job difficult.

Many tried to bribe their consciences in those days. I

have proofs of this in writing, but don't count on me to denounce any of the men and women who were struggling then with a problem that touches the very heart of human liberty and conscience. They thought otherwise, that's all. And that right, of which they always tried to deprive my father, has suffered too much for us to treat it with anything other than an infinite respect.

Afterward he had numerous troubles of various sorts; the litany would be tedious and, ultimately, incredible. From then on he would never again be invited to any international conference on genetics. He would undergo one financial investigation after another, but there was nothing to discover. Then the tax authorities challenged him about the deduction of professional expenses that is automatically authorized by contract for professors of medicine, on the grounds that "it includes the cost of transportation to the hospital, and it is a well-known fact that he travels by bicycle." He had to pay back taxes on a four-year period, but there was no fine, since he was in good faith and not to blame (the former tax inspectress was the one who had written, telling him to claim the deduction).

We have preserved a letter full of good humor that he wrote to the dean of the faculty in which he expresses his astonishment at having received neither a promotion nor a raise in salary in seventeen years! He admires the administration's constancy in his regard and understands that one must show preference to the researchers who have titles and projects much less important than his so as to encourage them.

Basically, he laughed at all that. What was really difficult for him was when the funding for his research was discontinued in 1982. The "mandarin" law forbidding professors of medicine from conducting research for more than twelve years was ultimately applied only to him and to three other unfortunate professors of medicine, who were then supposed to serve as an example. He lost his laboratory and the entire team that worked with him. The others, all of the others, thanks to their well-placed connections, managed to convince the university that their presence was indispensable to the functioning of the department.

He still drew a salary, but he no longer had office space, a lab, funding for research, or collaborators. That may seem incredible, but it is the truth.

Golden opportunities in the United States were offered to him; he hesitated and refused them. His heart was here, in the center of the Latin Quarter, in the France that he loved so much and that made such a sorry return. He found a new location on the *rue des Saints-Pères*, where he established the Institute for the Pre-born. He found ways of paying his research team. His international reputation in the scientific world brought him financial resources that he had not hoped for.

During the last fifteen years of his professional life, he would carry out his research projects thanks to funding from North America, England, New Zealand, and also the Institut Claude-Bernard. He traveled all over the world, lectured, and returned with awards, grants for the people on his team, or money for a research program. To be sure, he was never short of money, but it never bothered anyone that the man who continued to form future generations of French geneticists in his department at the hospital was obliged to go abroad begging for bread in order to continue his research.

Certain individuals would retort that it was because his

scientific investigations were no longer of any interest. A review of the titles of the projects that he worked on until a few months before his death is enough to silence such an accusation. Let us simply recall that his funding was discontinued at the time when he was interested in folic acid. The pretext given was that the research was useless. Besides, the vote was obtained thanks to the good offices of some of his closest friends and colleagues. In reality, the stances that he took were unsettling; it became necessary to remove him. He was working on the efficacy of folic acid in preventing spina bifida, for which there is now abundant proof, and on its beneficial effects in strengthening the fragile X chromosome. Today, without going into details, the usefulness of folic acid is recognized by the scientific community, and his research, taken up again by others, is considered to be extremely promising.

A few months before his death, he published with one of his female assistants a very interesting study on the connections between trisomy 21 and Alzheimer's disease, with findings that have since been confirmed. His last award-winning publications, his crowning achievement, would be concerned with cancer, of which he was to die in the following weeks.

Papa did not have a martyr complex. He hated to have people pity him. He certainly would have been displeased if the troubles that he went through had become known. That was not his concern. If I have alluded to them, it is simply to recall that, in a democracy like ours, "good folks don't like it when someone travels off the beaten track".

There were lots of people ready to defend human life, and they often had the best intentions, but sometimes certain

ones had forgotten that the primary duty is one of understanding and assisting. If one is too intent on being right,
one forgets that there are human dramas behind certain
actions of which one disapproves.

Meeting on a daily basis with mothers in distress,
Jérôme Lejeune quickly understood that it was necessary to help women so that they would not be forced
into abortion as their only alternative. Very quickly he
became the president of Secours aux futures mères (Help
for future mothers), which takes in women who are in
difficult situations as soon as they know that they are
pregnant. The Tom Thumb Houses shelter them from
the very beginning, if necessary, without waiting for the
seventh month, as is usually required by homes for pregnant women. When you are two months pregnant and do
not know where to go because the boyfriend or the father
wants nothing more to do with you, it is no use knowing that someone can take care of you at seven months.
Here, no one asks for identification papers, no one judges;
women are accepted and helped, so that they can make
a fresh start in life with their babies by allowing them to
be born.

He was also the scientific advisor to Laissez-les vivre
["Let them live"]. But from the 1980s on, the disagree-
ments with the president of the association became serious. The president wanted to turn the movement into
a political organization, under the influence of Jacques
Cheminade, a representative from the European Worker
Party who, as many will recall, has since then been a can-
didate in several elections in France.

This is the party of a certain Lyndon LaRouche, who
incidentally was doing a prison term in the United States.
Papa, who never got involved in politics, refused to leave
the door wide open for him to lead off those who, be-

cause of their prolife convictions, were disappointed with the Giscard administration and who might have listened to other siren songs.

He also cautioned his friends who were lawyers and doctors, who had some influence in various Catholic circles. In reality, behind the standard, reassuring remarks, Jacques Cheminade was concocting strange pacifist messages in which slogans from the far right joined with those from the far left.

When he saw that Jacques Cheminade was coopting the movement, he quit Laissez-les vivre, the excesses of which he did not appreciate. Thanks to his warnings, the Catholics did not fall into the seductive nets of Cheminade, who then failed to make his political debut in France. From then on he would be the target of the European Worker Party for years. Party members set up a curbside newsstand where they sold a paper called *Nouvelle Solidarité* that would make Papa out to be, in turn, "a lascivious viper"; "the Pope's assassin, with photos to prove it"; and "a KGB agent disguised as a practicing Catholic". We used to think it was rather droll, except when they came to our street to slash all four tires on our car or when, for several months, some youths who looked like they were from Eastern Europe kept us under a rude sort of surveillance from the street corner and followed us when we went out in the evening with friends.

We never really knew who these people were. The authorities came several times to investigate, and maybe they have confidential information about this movement in their files. This lasted for several years; I guess that they got tired of it, since intimidation wasn't working, and they looked for other victims.

Perhaps it is venturing too far to state that publication of *Nouvelle Solidarité* ceased the moment the Berlin Wall fell, for want of funds coming from the Communist East bloc. In any case, the future would prove that the European Worker Party was shamming conservative values in order to pave a political way for itself and for some especially dubious ambitions.

The battle over abortion left its mark on our adolescence. We were "Professor Lejeune's children". In certain places we were celebrities, while in others we were avoided like the plague. That quickly taught us that, since clothes don't make the man, we have to live with labels that don't define us.

But I must say that, even twenty years later, it still surprises me. After serving for several months as staff director in the governmental ministry for solidarity among the generations, my appointment met with much violent opposition, in spite of my best efforts. No one knew me, I had not yet said or done anything, but I had this original sin ignominiously inscribed on my forehead in bold letters: I was the daughter of Professor Jérôme Lejeune. The curious thing is that the most vehement attacks demanding my resignation came from those movements and from that sector of the press that claim to be in favor of tolerance and freedom of thought.

I wanted to tell them, "The crime that I have committed, in your eyes, is to have been born of the legitimate love of my father and my mother. In the final analysis, it is the color of my skin that you don't like."

It is quite difficult to clear yourself when you don't know what the charges are. If gaining favor in their sight requires denying my father, then they had better not count on me for that. For so much love given, for so much love received, what else is there to do but to bear witness?

Summers in Denmark

Long ago medicine made its decision,
and since then it has always fought
against sickness and against death,
for health and for life. For even
when nature condemns someone to
death, the physician's duty is not to
execute the sentence, but rather to
try to commute the penalty.

Every year the expedition began again. It had started in 1952 and was repeated each summer. Those were the heroic days of the 4CV. This automobile, purchased as it left the factory, had just barely four seats and no accessories at all. Still, it had a motor that could attain a downhill speed of 90 kilometers (55 miles) per hour and that had to be treated carefully so as not to stall on an incline.

The trip to Denmark lasted three days. At that time only Germany boasted a convenient highway system. Belgium's roads were surfaced entirely with paving stones, and the sign that appeared most frequently was the one indicating that the road was broken up. And then there were the children, who vomited as much as they could and put up with three days in an automobile with the patience one might expect from them.

Papa often told about the episode with the customs official. When the travelers arrived at the Belgian border, a zealous customs official told them to pull over onto the shoulder. No doubt he wanted to search this gypsy wagon that was evidently up to no good.

He walked toward them and initiated the procedures by demanding in an authoritative voice, "Your papers."

Papa opened the window and the smell, the horrible smell of the diapers and the vomit mixed together, wafted from the automobile. The customs official jumped back and yielded to the impossible.

"All right, drive on!"

Papa laughed heartily in telling this story, but what perseverance it took to endure these trips. He often said that it surpassed human patience.

With a grand philosophy and a lot of humor, he had written, moreover, a definitive statement on the happiness of being parents: "How difficult this very happiness is to put up with! Three charming children for two hours, and there I am on the verge of a temper tantrum. The wise books praise wisdom, and I reckon that any father of a family worthy of the name has to make it his chief virtue, to say nothing of his mainstay.

"Furthermore, life undertakes to prove to us a little more each day how much of a cardinal truth it is!"

He must have been encouraged by these thoughts during our infernal trips to Kerteminde.

Then there was the epic period. With the improvements to the roads and above all the purchase of a white 404 station wagon, which for us was a palace, the journey took only two days. Unfortunately, it was very soon impregnated with the fateful odor that was the stimulus that called us to action. The Dramamine tablets gave our parents something of a respite, at least up to the Belgian border. Then again, from two we had increased to five, and it was necessary to keep these little folks occupied during two days of travel.

That's not all. The 404 had a boat in tow. A Vaurien, which was the joy of Damien, my brother, all year long on the banks of the Marne River and which accompanied us to the Baltic Sea.

Before our departure we would spend several days in the country packing our bags. Mama made little pillows for each of us by putting our socks and shorts into little bags made of white linen. She also made sure to stock up on bonbons, which proved to be excellent elixirs for patience.

We used to put all the clothes out on the grass and then begin trying them on.

"You can still wear that this year. Give those pants to your brother; they don't fit you any more."

And that went on for hours. For us it heralded the departure, and we loved these madhouse moments when everything reminded us that we were going on vacation.

"Don't you think that this smells like Denmark?"

And we would speak French with a Danish accent, or else we would embellish our sentences with Danish words that made us laugh. "*Papelotte*", which means hair curler, was a big hit. We would wait impatiently for the day when our journey began.

"J Day" would start at four in the morning. There was the formidable ordeal with the Dramamine tablet, which I could never manage to swallow. Even drinking huge gulps of water would still leave it there on the tip of my tongue. It was awful! Papa would grow impatient and ended up crushing it in a little spoon and mixing it with some jam.

At five o'clock we were on the road. Papa set the odometer at zero, and we entrusted the voyage to the Blessed Virgin. There were suitcases everywhere. In the car, on the roof, and in the trunk. And la famille Fenouillard were off on their new adventures.

That evening we would put into port in Germany. Mama used to note in a little logbook the hours of arrival at each of our main stops and the address of the inn where we stayed, so as to be able to find it again the following year. It was wasted effort; we never found it again, and from 6:00 P.M. on we would start to explore the German countryside in search of a little hotel suited to the financial means of our parents.

Damien was already dreaming of the *Bratwürste* that were the requisite bill of fare at this stage of our journey, and Papa used to impress us with his German. Sometimes the innkeeper could not tell what he ordered. It was, no doubt, an "idiot" who spoke his local dialect and could not understand Papa's subtle accent.

The following day was the happiest of all. Each hour brought us closer to our final destination. We used to stop for lunch in Hamburg, at the home of Peter, a German student whom my parents had hosted for a time. There was a large garden, and the food was good. Peter did the dishes while telling us that he had learned from Papa the singular usefulness of this household chore.

"You see, it gets the dirt out from under your nails."

When we got back on the road, only a few hours remained. Once we arrived at Odense, no matter what the weather was like, we would open wide the windows. It was a game, who would be the first to catch the scent of the algae, the smell of the sea. Then we saw the fjord, the town, the bridge. We had arrived.

We would meet again with Farmor and Bestfar, our adoptive grandparents, who hoisted the Danish flag in their yard as part of our reception. In the yard there was a bush loaded with currants, which we would munch on while waiting for dinner. The meal was always the same: potatoes, *fricadelles* (meatballs filled with breadcrumbs and pieces of bread and seasoned with onion), and the traditional brown sauce that accompanies most Danish dishes and sticks to the ribs. In the land of Scandinavian mist and the Baltic Sea, the cooking is simple, unrefined, but natural and nourishing.

We had lodgings in several places in Kerteminde—always near the beach called Sydstrand, where Mama had always lived. They were always small apartments where, in order to bunk all the children, we would convert the cellar into a dormitory. It was rustic, a little humid and dark, but to get to sleep in the cellar was for us the sign that we were grown up now.

Denmark, the site of all our summer vacations until we reached the age of eighteen, was our favorite place, where we established immutable traditions that we were delighted to rediscover each summer. This tradition still goes on, and at one end of the beach, which Mama and her childhood friends had staked out for themselves thanks to their numbers and their persistence, all the friends in the neighborhood gather, now with their children and grandchildren as well. Pity the unfortunate soul who would try to ensconce himself on our *gryde*. The beach was a wilderness that had been invaded by rank weeds. The stretches of fine sand, protected from the wind by the vegetation, were called *grydes*. Someone would first have to get up at five in the morning to arrive there before Mama, who, with the dawn, goes to take her morning bath in the Baltic Sea and spreads out blankets so as to dissuade eventual invaders.

Generally, folks in that locality know better and do not dare to venture onto territory that is so well guarded, but it has happened that poor visiting strangers have taken the risk of placing their beach towels a few meters away from ours. They packed off pretty fast, poor people, when they saw themselves overrun by hour after hour of grandmothers marshalling their children and their grandchildren, whose mission was to make as much noise as possible and to launch poorly aimed beach balls, which they would go and retrieve politely, saying "*Pardon, madame*", while putting their sandy wet feet on the towel that the lady, coated with tanning lotion, had carefully spread out a few seconds earlier.

For forty years this routine had not been challenged successfully, not even during the dog days that occurred once every decade or so, when factories closed and children were sent home from school to take advantage of a blazing sun that had taken a wrong turn for a couple of days.

Of course, you must admit that the best allies of these ladies who are seeking tranquility are time and tide. It may rain, the winds may blow, and the sea may be let loose, but nothing will ever interrupt the unchangeable ritual of meals on the beach and get-togethers at three, the *cafetid*, when even those who are working take a break for a quarter of an hour, jump on their bicycles and come to drink coffee and eat *brunsvir*, a cake made with leavened dough and covered with brown sugar, butter, and almonds. A very nutritious snack after an invigorating swim in the Baltic.

Marcelle Lermat, mother of Jérôme Lejeune.

Pierre Lejeune, called
BonPa, his father.

Jérôme Lejeune (left) with
his brother Philippe.

Jérôme Lejeune (left)
playing Arkel in
Pelléas et Mélisande
by Maeterlinck.

Jérôme Lejeune at
the age of thirteen.
First Holy Communion.

Birthe Bringsted at age twenty-two. Danish citizen. Wife of Jérôme Lejeune.

Jérôme Lejeune and Birthe Bringsted. Newlyweds in Denmark.

Jérôme Lejeune, Sunday gardener.

Jérôme Lejeune and his fourth
daughter, Clara, in 1964.

Birthe and Jérôme Lejeune with their
five children in 1971.

Professor Lejeune and his wife being received by John Paul II.

Jérôme Lejeune, appointed to the first professorial
chair of fundamental genetics in France in 1964.

Vacation in Denmark, summer of 1968.

Jérôme Lejeune and his research team. Neckar Hospital, Paris, Pediatrics.

Lejeune family reunion.
Christmas, 1993 a few months before the death of Jérôme.

Jérôme Lejeune with
his "dear little ones".

Even if they don't want to admit it, Mama and her childhood friends have become thoroughly bourgeois. When we were little, we would be set up at the edge of the beach in all sorts of weather, and when it rained our mothers would pile us into a little blue tent. That could last for days on end, and we would pass the time by eating little cookies that were called *Marie-Kiks*. The game was to grab them between two toes and to eat them without dropping them and without using your hands.

What's more, we used to have carte blanche to eat as many as we wanted, since they were not expensive and there was a contest to boot. On every package there was a round label in the same shape as the cookies, divided into quarters like camembert cheese, each of a different color. You had to collect a certain number of different colors to win. I think that we ate hundreds and hundreds of them over the years, but our mothers could never complete the set. Yet we must have been the biggest consumers on the isle of Fionie. Some years later, fed up with *Marie-Kiks*, we sulked when the only cookies we were allowed to have were these biscuits that were more like hard tack, and our mothers concluded that the contest was rigged. It was impossible to win, the only winner being the factory that made these dry cakes, which had managed in this way to sell off its merchandise. The *Marie-Kiks* still exist, but they certainly no longer play a leading role in the Kerteminde market.

In 1968 began the Age of Plastic. It made its way onto our beach with the ingenious invention of a bowl-shaped tent made of transparent plastic through which, as the advertising claimed, one could get a tan. And so the ladies were ensconced in their "bubbles of light", taking advantage of the faintest ray of the sun, even when the temperature outside was practically inclement.

In 1975 we experienced the Copernican Revolution. Mama acquired a small cabana (with twelve square meters of floor space) on the beach, and we abandoned thirty summers in common on our *gryde* to move a hundred meters down the same beach to the other side of the ice cream stand. That had always seemed like the distant outskirts, but, to our great surprise, we got used to the comfort of this wooden hut on the pebbly beach.

Such a break in millenarian traditions was not made without a period of transition. At first Mama moved her blankets, her umbrellas, her coffee, and her famous *brunsvir* in an old perambulator that had once served as a naptime crib for all the children on the beach. Passersby would see her pushing her baby carriage, filled to the brim, on the "promenade" that ran along the beach, and a few of the braver souls bent down to look at the poor baby hidden under all that cargo. Occasionally they would see a pork roast, some smoked mackerel, or a whole assortment for supper. It depended on the hour.

So it was decided that the cabana—the *badehus*, to use its real name—would serve as a refuge "in case of bad weather" and would allow the "ladies" to leave their things at night instead of having to bring them home on their bicycles. It seemed impossible to give up our magnificent *gryde*.

As it turned out, for several years in a row the "in case of bad weather" occurred more often than it was supposed to. Headquarters moved to the *badehus* after all, definitively. There was something else that worried Mama a lot. With two settlements it could happen that less determined visitors who came to see Mama on the beach would go away disappointed, thinking that the bad weather had won out over her tenacity. They seriously underestimated her, though. Still, it is true that with only one base it was no longer possible to make that mistake.

Don't let that mislead you. Though we did not like intruders, all the friends of our friends were welcome. Mama always did cook for an army, and on a small portable gas stove, suitable for expeditions on the northern face of Mount Everest, she managed to prepare sumptuous meals for all the friends who joined us.

And Papa?

Papa did not like the beach, or the sand, or the gabbing, and he liked even less the idleness. But he liked Mama to be happy, and to see her gossip with her girlfriends from school and relive a bit of her childhood made him happy.

He came each year to travel with us and again to take us back. He stayed a few days and then left for a solitary summer to continue his research in the stifling heat of Parisian summers. He suffered much as a result of this separation but said nothing about it. Every evening he sat at his desk and wrote to Mama. For their whole life, they both wrote to each other every day, faithfully, tenderly, when they were apart. The letters that my parents exchanged during all the summers of their life together, still, reveal a weariness with being alone.

In Paris, the only one at his office, he constructed scientific models, wrote his lectures, rethought molecular biology, and, in flashes of insight, comprehended several of the subtle mechanisms of the human mind.

It was in the suffering of his isolation that his labors were most productive. The telephone didn't ring any more, the patients were gone, the laboratory was deserted. The mind, freed from everyday concerns, enjoyed a breath of fresh air: the perfect opportunity to exercise its reasoning and deductive skills; the prudence of a scientist who tests his working hypotheses a thousand times rather than just once; the virtuosity of an intellect supported by scientific, medical, and biological knowledge that has always impressed those who have encountered it.

He also took advantage of the solitude to travel and to lecture just about everywhere in the world. Photographs, tossed at random into archive boxes, today recall his many voyages around the planet to explain, to announce findings, to convince, and to learn, too, untiringly.

In Denmark, Papa used to rediscover nature. He would go exploring in the forest or stay and work in the deserted house, immersed in his eclectic reading. He offered sacrifice every day at the *cafetid* ritual and, with his customary courtesy, chatted and joked with the friends who were present.

He also did some experiments. He tested the elementary concepts of physics and fluid mechanics by taking up a position in a boat and gesticulating enough to sink it. For several years he took advantage of these short moments on the beach to work in flint and to reconstruct the entire toolkit of prehistoric men. He taught us how our ancestors lit fires, carved knives, and made axes. . . . He also tried his hand at drawing, and we have found sketches that he made of the bay of Kerteminde or of his grandchildren.

One day when the weather was fine, Papa was pulling through the water a child seated in a beach boat. Friends of mine called me over to tell me that there were some people asking for me. I saw two fellows and a girl. They were a bit ragged; they were twenty years old. After a moment of hesitation I recognized one of them. I had seen him on only one occasion at a party hosted by some friends—a friend of a friend who knew about my Danish origins and about the legendary hospitality of Mama. They had just made a beautiful but rough trip through the great Scandinavian north and had reached our corner of the beach, guided by people from the village. Without saying so, they were hoping for a good shower, a place to stay, and lodging.

A bit flabbergasted, I invited them to come see my parents on the beach, and suddenly I saw the big dark-haired guy whom I didn't know and the girl stop in astonishment, turn to each other, and say, "Can you believe that that's him?"

Those two, who since have become our great friends, had on several occasions attended lectures given by Papa. They had admired his courage and supported his positions against abortion. They had seen the great scientist and the orator, and now they found him on this pebbly beach on the Baltic Sea, running through the water, pulling after him a child in a boat and laughing heartily with the tyke.

To tell the truth, what could be more normal? The astonishment of our friends surprised me.

Then one day the storm would come. The sea became rough, and you could feel the tide. The sky turned black, and the wind blew everything away. The rain of the equinox announced that the vacation was over. Our little friends had already returned to school several weeks before. We had to go back to Paris.

Mama would get our bags ready and put in them a little bit of Denmark for the winter: some red cabbage for Christmas, some herring and smoked meat for her guests, some *coulor* to make good brown sauce, a little black bread and feather bedding for the children. We would say goodbye to Farmor and Bestefar, our adopted grandparents, who then hugged us very tight, with tears in their eyes, always worried that they might not be there the following year and that they were seeing us for the last time.

We would drink our last bowl of milk with our friends, in the evening after dinner, in the shed at the back of the yard, which we had turned into a HQ and where we used to meet every evening with the children from the town. After the stroll around the port, our last look at the sea, we would return, very sad, to go to bed. Next year is so far off!

In the morning, at dawn, a few brave friends were waiting for us. They had jumped onto their bicycles to say one last goodbye to us. With the boat hitched to it, the car drove off and turned its back to the fjord, to the sea, and to the memories. I scarcely remember anything of the return trip. It must have been just as long as the one going there, though, but less charged with emotion and anticipation. The only image that comes to me is that of the crowd on the Boulevard Saint-Michel, the smell of the Rue Galande, the pleated navy blue skirt, the white blouse, and the school apron, which were ready for the start of classes. I remember also how difficult it was for us to speak and to think in French for two or three days, to do without the red wooden shoes, and to have dinner so late, just before going to bed.

A Witness in Our Times

The one who first knew that he would have to die and then constructed tombs, the one who rescued his wounded fellow creature, took care of him, and protected him in his weakness through long years, as fossils prove to us, the one who discovered art that went well beyond mere technique, that one who is one of us, not even a hundred thousand years old, possesses something like a spark of intelligent love.

In 1957 Jérôme was the young father of three children. For the first time he left his family to "go to America", as they used to say in those days. America had a cachet of adventure and conquest that the term *USA* could never convey. For a research scientist with the soul of a poet, this nuance had its significance.

He had just been named as the expert from France to the United Nations' scientific committee on the effects of atomic radiation. Some years after Hiroshima and Nagasaki, the formation of an international committee bringing together renowned scientists, coming from all over the world to deal with the agonizing question of nuclear power, was truly an event. For the first time, Americans and Soviets, in the middle of the Cold War, would find themselves at the same table and attempt to compose a joint document.

He wrote from the Hotel Windsor about his first impressions of a cosmopolitan, passionate city that gravitated around the United Nations. It was the start of a long correspondence by aerogram that my parents would carry on with each trip abroad.

There he met all the big names in medical and physical science. He forged bonds of friendship with the most renowned geneticists of the day, all of them Americans. France was only taking its first steps in this brand-new specialized field.

So began, with an American debut, the international career of this "Frenchie", who in a few months learned English by the Assimil method in the evenings, after the sessions. Very quickly he acquired a perfect command of English, but with a delightful accent that the Americans found melodious.

It is thanks to this committee that France banned those magic boxes that used to be in the shoe stores. You would put your foot on it, and, thanks to X rays, it was easy to see whether your toes were properly aligned or cramped. The dose of radiation, which might be acceptable if you only bought one pair of shoes per year, nevertheless put the shoe saleslady at a very high risk.

Beyond the results alluded to in this anecdote, the essential concern of this committee was to bring together scientists to discuss a subject that was politically very delicate, at a time when the Americans and the Russians had launched into an unprecedented arms race. On several occasions Jérôme, the reporter for the group, found a mutually acceptable wording that satisfied the Americans while allowing the Russians to save face.

It would require of him a lot of talent and strength of conviction in the midst of the Cold War, when the missiles of Cuba during the Bay of Pigs episode and the hot spots in Asia and Africa were amplifying the rumors of a third world war, which was in great danger of becoming a nuclear conflict. But the experience would win him the real friendship of American scientists and the respect of his Soviet counterparts, who would be grateful to him for not locking them into a sterile debate predetermined by political exigencies. The little Frenchman thus played, in his sphere, the role of a mediator, demonstrating in his own way the role that France hoped to play on the international scene.

In 1962 he was named expert in human genetics at the World Organization of Scientists. From then on his international reputation was made; he was the first to discover a sickness caused by a chromosomal aberration and thus opened a new field of research. On several occasions he would be called as an expert witness before national parliaments, commissions, and courts in the United States, but also in Austria, in England, in New Zealand, and even in Moscow.

In those days the little suitcase with wheels did not yet exist. He had invented a prototype that allowed him to travel with a carry-on bag that could be rolled through the labyrinths at the airports. The execution of his design was not as aesthetically pleasing as the finished product that would be found later in the leather goods stores, but the model was extremely durable and often elicited the interest of his traveling companions, who were ruining their backs with heavy luggage.

As a pilgrim he used to have his travel kit for the airplane flights. An old sweater that was very warm and a shapeless pair of pants so that he could doze off more easily, some sleeping tablets, a half tablet of Imménoctal to prevent the discomforts of jet lag, and a boxwood decade rosary, which he had made one peaceful Sunday in the country.

In the middle of the papers that he had compiled for his deposition in the [Blount County] Circuit Court at Maryville, Tennessee, in the United States, Jérôme is carving wood. From a little branch cut into segments he is making little oval-shaped objects. The center is hollowed out according to the dimension of the future owner's index finger—the size of the first phalanx, (that is, the tip), but the second (joint) must not go through; that would be too large. Around the outside there are pretty little teeth, sculpted and then sanded down so that they are smooth to the touch and produce a calming sensation. At the top of the oval, the tooth is larger, and a cross is engraved upon it.

My father carved hundreds of decade rosaries in this way. For him it was the happy opportunity to enjoy manual work, to let his mind wander and thus reflect more freely, and to pray joyfully as a laborer who offers to the Lord the work of his hands.

His wooden decade rosaries were a great success. They were beautiful, they could be kept in the depths of a pocket or in the hollow of the hand, and each one was unique. In times of happiness or of grief, they accompanied many whose names are unfamiliar, but also many well-known people who had requested them from my father. In this way they carried with them a little of his friendship, his courage, and his prayer.

The fortunate recipient of this present had a guarantee: not that he would never lose it, but that he would not be overcharged. The only compensation asked of him was to recite a decade per year for the woodcarver.

Jean-Thomas is fourteen years old. He likes to visit his grandfather in his office. Like all grandchildren, he finds a pretext or a question to ask, which gives him a reason to knock on the door. He likes so much to see him working with wood, and maybe he will let him try?

"Grandfather, is the weather cold in Moscow?"

"In this season there is sure to be snow, with temperatures of minus forty degrees. When winter sets in there you can't go outside without a very thick coat, gloves, and a warm fur cap."

"Have you been to Red Square?"

"Well, yes, I've shivered in the cold there several times; you see, I'm a regular at presidential funerals. The Holy Father sends me to attend them as a representative of the Pontifical Academy of Sciences. And since their leaders are very old, they die often. Moscow is a magnificent city, even if the sadness and the misery are as plain as day. The great soul of Russia is still alive, despite the Communist dictatorship."

"You talk about friends over there. Haven't you ever been threatened by the KGB?"

Grandfather smiles and puts his arm around Erwan, his nine-year-old grandson, who is curious about everything and loves adventure. At that time the Berlin Wall has not yet fallen, relations with the West have not thawed, and communism still weighs heavily on the East.

"No, my boy; I even have some real friends there."

"But that can't be; you're against the Communists. And there's plenty of them over there. Weren't you ever afraid that they would kidnap you?"

"I've been traveling to Russia for quite a long time now. You see, the first time was in 1964. I had to give a lecture for the Soviet Academy of Sciences and speak of modern genetics. The only thing was, at that time there was an official doctrine, even for a science like genetics that has so little to do with politics. A certain Mr. Lysenko, who claimed to be a scientist, had constructed a theory, according to which an acquired characteristic becomes innate in the course of several generations."

"What does that mean, Grandfather?"

"It's very simple: everything good or bad that you learn in your life changes you genetically, and then you transmit it to your children. It was scientific nonsense, but very convenient for the Communist authorities. A good Communist had children who were good Communists. In this way they were building a better world, and, incidentally, this allowed the children of the men in power to have privileges going far beyond the difficult daily routine of most people."

"But that's stupid. Any idiot can tell that's not possible?!"

"You are right, Jean-Thomas, but it was said in a more subtle way with scholarly words that confuse everything, and don't forget that it was dangerous for the Soviet scientists to say anything different."

"Well, then, what happened?" asked Erwan impatiently. "Did the KGB try to put you in jail?"

"No, but when I wanted to give the lecture—which, of course, was opposed to 'Lysenkoism'—the great hall of the Academy of Sciences was officially unavailable because of urgent repairs started the day before. It was a way to prevent me from meeting my Soviet colleagues and to make me take the first plane back to Paris.

"At that time there was a World's Fair in Moscow. I met the French ambassador and asked him whether it was possible to lend me a room in the French pavilion. He welcomed me with a great deal of courtesy and permitted me to give my lecture.

"At the start, in the midst of the audience from the French community, I saw some Muscovites, dressed in thick coats with the collars up and their caps down to their eyes. Yet it did not seem so cold to me. As I went on speaking I saw them notice each other. One Muscovite took off his wraps first, and the others followed. They were the members of the Academy of Sciences. They had come, they had counted how many of them were there, and they had seen that they were in the majority. Some time afterward, Lysenko was overthrown by a vote of the academy. That is why I have real friends in Russia."

"But the KGB, didn't they follow you?" asked Erwan, a little disappointed that Grandfather was not considered a dangerous spy.

"In those days someone would always accompany you, wherever you went. The tourist guides were really spies who kept us under close surveillance."

Wow! Erwan is relieved. Grandfather is a hero who stood up to the KGB!

"They even tried to trap me with a young woman. She was very beautiful and couldn't take her eyes off me. But it was a rather crude setup, and you can't teach an old dog like me new tricks. How could she be interested in an old man? I found their plot terribly unimaginative, or maybe it was just a routine thing."

Jean-Thomas finds that Erwan is getting tiresome with his spies and his KGB.

"Did you go back to the Soviet Union, Grandfather?"

"Yes, several times, but the most remarkable trip was in 1981, when I went to Moscow to see Leonid Brezhnev, the president at that time, to bring him a message from the Pope."

"And the spies let you, Grandfather?" Erwan exclaims.

"What in the world did the Pope have to say to him that he would send you with the message, Grandfather?" Jean-Thomas asks in astonishment.

Grandfather set aside his woodcarving long ago. With his fingers he caresses the decade rosary that he was working on, as though to polish it smoother. The two boys listen, sitting on either side and leaning against him, as we used to do as children when he returned from his travels.

"If you're interested, I'll tell you the whole story. Since 1974 I have been a member of the Pontifical Academy of Sciences. It is an academy that brings together scientists from all over the world and informs the Holy Father about the findings and the development of all branches of science.

"When I was young I was an expert at the UN on atomic radiation. Because of this position I participated in a working group of the Academy that was supposed to study the consequences of the proliferation and the increasing sophistication of atomic armaments. We had reached the conclusion that in the future only the wisdom of men could prevent a nuclear war. Because such weapons have been miniaturized and made more widely available, the great nuclear powers would be longer be prevented from having recourse to using them, should they be participants in a conflict.

"And so the Pope sent an emissary to the heads of state in all the nations that had the atomic bomb, to bring a message of peace and wisdom."

"And did you go to all the countries?"

"No, only to the major powers that possess nuclear arms: the United States, Great Britain, France, the USSR and China. I was sent to see Brezhnev, who welcomed me with a great deal of courtesy.

"At that time the Russians and the Americans were conducting an incredible technological war to become the leading nuclear power. It was an arms race that eventually bled the Soviet economy white.

"As unbending as justice, sick and almost crippled, Brezhnev received me with unbelievable pomp in a ceremony worthy of the tsars. On his impassive face I saw a glimmer of agreement when I read the Pope's message to him. Yes, only human wisdom could prevent the massacre; only the intelligence of the great ones of this world could stop this arms race which, by its bellicose logic, could not help but lead to catastrophe. After a formal speech that he read to me, he pronounced words of peace, like a man who aspires for rest. That greatly impressed me."

"And do you know a lot of heads of state like that?"

"You know, my boy, no man is a hero to his valet. But it's true, I have met many of them, face to face, who wanted to understand a little better what science teaches us about the origins of man. It is a great mystery that interests everyone, even powerful men. I have seen the Emperor of Japan, the King of Spain, the King of Belgium, the President of the United States, the Queen of England, and others as well. I have also testified before foreign parliaments, like the English House of Commons and the American Senate. To them all, I have always explained the same story of Tom Thumb, which you know well, the most beautiful story of humanity, which renews itself at every moment in this prodigious miracle of a life beginning in secret."

"And how did you do that?"

"I would speak to them just like I'm speaking to you. But the most beautiful experience was the trial in Maryville. I even made a book about it, *The Concentration Can* [San Francisco: Ignatius Press, 1992]. I will tell you the story the way I have done for many audiences."

It occurred during the week of August 15. I happened to be in the laboratory, and I received a telephone call from one of my American friends, Mr. Palmer, who is a lawyer. He told me, "While reading the newspaper today I learned that at Maryville, in Tennessee, a judge is hearing a case during the next few days. It's a trial in divorce court: a young woman, Mary, was sterile because her fallopian tubes were completely obstructed. They tried artificial insemination with her, but it didn't work; then they performed an in vitro fertilization." That is, they removed several ova and fertilized them in a test tube.

Two embryos were implanted in her uterus, but unfortunately they didn't survive. And the other seven were put into a deep freeze. My friend Palmer then told me, "It's a trial in divorce court. Apparently the couple could not cope with either this new sorrow of two lost infants who did not come to term or the idea of having their own children in a deep freeze. As a result, they're going through a divorce. They are in total agreement about dividing all the property: the apartment, the car, but not about the children. The husband insists that they be destroyed; the wife asks that custody of them be given to her so that she can bring them into the world. Do you want to come and testify in Maryville?"

I said, "Testify about what?"

"Testify that these are human beings, because in American law, in matters of divorce, there are only two categories: either you're dealing with patrimonial goods that can be liquidated, or else it's a question of children who must be assigned to a guardian."

Providence sometimes gives signs. I then asked him, "But really, what does the young woman say?"

"Well, she said, 'If the Court does not let me raise my children—she called these little frozen embryos her children—I still don't want them to be killed. If they don't want to give them to me, at least let them live by giving them to another mother.'"

So I said to Palmer, "All right, if that's the case, then I'll come. The trial has already been decided. That woman answered in exactly the same way as the real mother answered Solomon. You know, the judgment of Solomon is pronounced, as far as I know, about once every three thousand years; if it occurs during your lifetime, it's worth a detour."

So I went to Maryville. The story, though, is scarcely believable, and I'm telling you nothing but the honest truth. This woman's name was Mary. The trial took place in Maryville, and her lawyer's name was Christenberry. I didn't know that until I had arrived there.

The sequence of events was absolutely extraordinary. This trial caused a stir throughout America for one reason, to be precise: it was the month of August, and the Loch Ness monster had not yet reappeared.

So it was page-one material for American newspapers,

and every day they carried photos of Mary and of her husband, and of me, too.

The amazing thing is this: there were about fifteen cameras there that transmitted everything to satellites, which then broadcast it all over America. Curiously enough, I had the impression that these people were not taking it lightly. It was really something whose importance they appreciated.

I will not repeat to you the testimony. It was very simple. It was a question of explaining, as a geneticist, that it is well known that sufficient information—all that is necessary—was there at the moment of conception, and that there was no doubt that these were very young human beings. Extremely young. Incredibly young. But they were living beings, and their biological inheritance allowed us to declare that they were human. And a being that is human is a human being.

The fact that they had been preserved in suspended animation (in lowering the temperature, one stops the movement of the molecules, and finally time itself stops) made no difference in the situation: time was suspended for them, but, if one were to give time back to them, they would revive.

Therefore, in order to make the judge understand what was at stake, I used a very simple word. I told him, "These very young human beings are frozen, packed together by the thousands into an extremely restricted space, where time itself is stopped. They are confined, to put it exactly, in a *concentration can*."

Certain journalists in France translated this as "concentration camp". The alliterative allusion was perhaps not unintentional; nevertheless the translation was erroneous. For a concentration camp is a device for accelerating death, while a *concentration can* is a device for slowing down life.

In both cases, innocents are imprisoned, and probably

the innocence of those being held in concentration caused the journalists to confuse camp and *can*. I fear that our society is just as seriously confused about can and camp and that, when it is a matter of technology, people accept this concentration of innocents, even though it is no more innocent than the concentration brought about in the past in places that were on a vastly larger scale, but where men, too, were concentrated in a glacial setting, where not only they themselves were arrested, but their future as well.

The judge thought it over for a long time. He said that he would hand down a judgment at the end of one month. He wrote a forty-page ruling/opinion. I have read it carefully—it is absolutely extraordinary.

Dale Young, an unknown judge in a tiny jurisdiction, Maryville, near Knoxville in Tennessee, has written definitively for American common law that men begin at fertilization. The judge gave judgment the way Solomon had judged. He said, "The person who should have custody of the children is the person who intends to preserve the life of the children."

Jean-Thomas and Erwan listened religiously. They were more grown up when they left, because their grandfather had confided in them as adults in teaching them about the book of life. Grandfather returned to his woodcarving and now is sculpting, like a prayer, the next decade rosary, which he has promised to give to a young priest who is beginning his apostolate in a difficult suburban area.

Sunday in the Country

> *For the geneticist, it is quite*
> *remarkable that we use the same*
> *word to denote an idea that comes to*
> *our mind and a new human being*
> *who comes to life: conception. One*
> *conceives an idea; one conceives a*
> *child. And the science of genetics*
> *tells us that we are not wrong in*
> *using the same word. What is*
> *conception? In reality it is the*
> *information inscribed in matter,*
> *so well that this matter is no*
> *longer matter, but a new man.*

Oratorical jousting has nothing to do with combat. No rivalry, no concern with dazzling the interlocutor or of subduing him with one's own intelligence and talent.

The two men are back from the front lines. They discuss and they debate while contemplating the universe. Each one enriches the other.

It is a flamboyant display of intelligence, humor, feigned anger, and, occasionally, gross exaggeration. The stream of words determines the course of the thought. Saint Thomas Aquinas flirts with the new invention of the laser disk; the speakers take offense at the latest edition of Pivot's talk show, when So-and-so, the last person in the world to ask about art, said nonsensical things about pre-Raphaelite painting; they look at the late-model car just bought by the one while discussing the latest scientific research of the other.

The one paints the souls; the other treats the bodies.

The one is an intellectual; the other is an artist.

The one leads a monastic life; the other travels all over the world giving lectures.

The one lives apart from the world; the other lives in the world.

The one contemplates the world in its creative splendor; the other studies it and understands it in the intimate and intricate machinery of life.

Both of them are men of faith and conviction.

Both of them have an immense talent.

They are united by a true love as friends.

They are brothers; the bond between them is indescribable and sacred.

Uncle Philippe is the older brother of Papa. He is an artist, a painter, but also a philosopher, a theologian, a music lover, and an ingenious handyman to boot. He calls himself an autodidact, and his amazing, eclectic education proves it.

A man of passion and of self-denial, he lives withdrawn from the world and lives intensely the life of the world. He founded a studio for the young people in the town of Etampes, which, in the course of time, has actually become a school of painting.

Everything fills him with enthusiasm, enchants him, exasperates him, irritates him prodigiously, or else leaves him completely indifferent. He is the opposite of lukewarm. He practices his faith as his reason for being. He carries his cross, and he is radiant.

Like his brother, he is a messenger, one whom the modern age does not understand. He paints what is essential in man and in life: the brightness of colors and of movement, illuminated by the divine Light. He says to his students, "Your work must not be viewed for itself, but for what it shows beyond itself."

Indescribable himself, who else can better fathom the indescribable, by his regard for beauty and his contemplation of God?

Uncle Philippe seemed to us to be the same age at seventy as he was at thirty. His features are a mixture of the old age of grief and the youth of hope, true hope. Every Sunday, at around four in the afternoon, his emaciated and slightly stooped silhouette walks up to the house. He comes to converse with his "old brother", as the two of them call each other. Once we reached adolescence, we would sit in with delight on this conversation, which consisted of a little bit of this and a little bit of that, of bland current events and of philosophical and scientific observations, moments of silence and flashes of humor. Oh, without a sense of humor, how would they ever have survived?

They have this strange idiosyncrasy of always seeking an explanation for what they have seen or heard. The most incongruous hypotheses are sometimes proposed, but that does not shock anyone. They are simply chatting; maybe they will mention a contrary argument the next minute. The train of thought moves freely; it seems to be touring from place to place in their memory and their knowledge. It is sometimes a bit much, sometimes wrong, sometimes right, often funny, always fascinating.

These two men, who have always sought to understand the heart of our humanity, have never thought it

worthwhile to reexamine critically the spiritual and intellectual patrimony that their parents had entrusted to
them. Uncle Philippe said to me one day, "Never forget
the essential role that my parents played for their children. They were both highly cultured; they loved music
and painting. They gave us a taste for literature, and by
the age of fifteen we had become acquainted with all the
great classics and loved them."

As adults, they considered that what their parents had
given them was beautiful and good. They cultivated it
without contesting it.

There was never any rivalry or competition in their
relationship. It was a strange thing, but every time one
of them had made a discovery, completed an important
project, or simply found a line of investigation to pursue,
he had to bring it to the other one for his criticism before
he would consider it valid.

Papa's family could provide the framework for a novel;
there is a bit of Balzac in the social conformity of their
behavior and the dramatic intensity of their feelings. The
only thing we know about them is their pictures from
Epinal. There was a great-grandfather who sold orthope-
dic shoes at a store near the *Place de l'Odéon*, and a veterinarian grandfather who had invented bovine ovarectomy
and composed the song, "A la volaille", which, it seems,
is still sung by the veterinary students of Maisons-Alfort.
He was accused by his in-laws of reading Voltaire and
being a "free thinker", but he came to no harm, thanks
to his very pious wife. They only had one daughter, Mar-
guerite, Marcelle Lermat, who received the best possible
education and would become my grandmother.

At the age of seventeen she married Pierre Lejeune, son

of the mayor of Montrouge, who was ten years older than she. He surely married her for her beauty, which combined a rounded figure with an unblemished face, bright eyes that were slightly protruding, with a very fine mind and a vivacious temperament. One month later he went off to war, and this very young wife set up house alone in Paris. In those days, even at the age of seventeen, it was inconceivable that a married woman would return to live with her parents. In 1915 she welcomed her husband home on furlough, muddy, stinking, covered with dust. She embraced him and undressed him at the threshold, while a hot bath was running.

Pierre and Marcelle would have three children, after ten years of waiting: Philippe and Jérôme, who were raised together, and Rémy, who would be born ten years later. Every summer Marcelle used to leave with her sons for Royan, in a Ford V8 that she drove herself, to visit her parents.

But the childhood memories of Papa and Uncle Philippe are marked most of all by the war. They were fourteen and fifteen years old in 1940, and from their observation post on the wall of the property of Etampes, they used to await the arrival of the Germans with a very lively curiosity. The war was a great adventure; they were no longer going to school at Stanislas, they were reading the classics, reciting verses of Aeschylus, and learning to "enjoy their leisure", the expression of Montesquieu that Uncle Philippe loved to quote.

A German on a bicycle, with a big gun strapped to his back: that was the first glimpse that they would have of the German invasion. There was nothing to conquer; everybody was gone. The Communist mayor and his town hall staff had fled; the populace chose to make an exodus. Pierre, their father, considered that there would probably be no combat and decided that it was preferable to stay.

Immediately the house was requisitioned so as to make it into a field hospital. The "medical officer" moved into the two rooms of the maid's quarters, and the injured men poured in.

The memories that they would preserve of the war, mixed up with youth and innocence, are astounding to us. They were cold and terribly hungry, since my grandmother found it utterly beneath them and disloyal to have recourse to the black market. Their father, by popular demand, was mayor of Etampes during the entire war. He spared the town and its inhabitants the extortions of the Germans.

One day he received a visit from a friend who proposed that he join the Resistance. He refused, because he thought that his role was to stay there, with the people of Etampes who were not able to leave; he did, though, hide the fortune of a Jewish friend, obtain ration cards for those traveling through, and organize lodgings for British pilots. One day the Germans asked to dismantle the statue of a marshal of the French empire so as to make ammunition out of it. With a cunning smile, Pierre Lejeune told them, "Go right ahead." Covered with a mold resembling verdigris, the statue appeared to be made of bronze. It was stone. At the Liberation, because he had remained as mayor of Etampes, he was imprisoned for five months and then set free because of insufficient grounds. Convinced that he had only done his duty, he emerged a broken man, even though that imprisonment surely saved his life.

The reason was that, after France was liberated, the family discovered the horror of the concentration camps, and all of their certitudes crumbled. An atrocious world opened up before their eyes, and nothing would ever be the same.

Every Sunday until she died, Mamie would come with her son Philippe to pay us a visit in the country. Mamie didn't like us to step on her toes or to kiss her with our hair in our eyes. She loved the cinema, witty jokes, and chocolates.

When we were a little older she would teach us to solve crossword puzzles, at which she was prodigiously talented. We would discover her fine, subtle sense of humor and the incredible originality of mind belonging to a woman who remained attached to social conventions. She would never even hint that she had defended her husband heroically and with dignity at the Liberation and that she had gone through the entire war, in occupied Etampes, with head held high and without faltering, in the face of the Germans.

I did not know BonPa, my grandfather. He died in 1958, no doubt of the same disease as his son, which at that time was not well known. I know that he loved Mama very much. I also know that he ran a little distillery that he had inherited from his father, but that the business side of it, which he managed satisfactorily, annoyed him a lot. He thought that his duty as a father was to pass on to his children a little of the culture of a gentleman. He had them recite verses from Aesop and Aeschylus on the way to school; one of the vestiges that we have today is the story of that stupid donkey, laden with sponges, who made fun of his companion who was loaded down with salt and laboring under the weight of it. They crossed a river, and it was the other one's turn to tease.

Most importantly, he made a decision that would prove essential in the development of these two inseparable adolescents. In 1941 he decided that his sons would no longer go to school. He gave them several Latin translations to do and otherwise guided them in their choice of readings. Philippe and Jérôme read all the classics, knew thousands of verses, and were fascinated by the universe created by Balzac. They laughed heartily at the jests of Rabelais and read Latin and Greek without difficulty. They also launched a grand enterprise: founding a theater company. In the freezing cold winter of 1941, with their friends, they met in the workshop opposite the house and organized a production: they made scenery for each play and planned performances in the nearby villages, transporting their costumes and the scenery on modified pushcarts behind their bicycles. They had some success—it appears. Seduced, moreover, by these brilliant beginnings, one of their friends would continue and become a professional actor.

Papa often spoke about this period, in which the world was opened up to him through reading and the arts. Through puttering around, too, because in those days when everything was scarce, when someone who had a nail was rich, it required a wealth of ingenuity to give shape to what the mind had invented. Their penchant for experimenting, moreover, went so far that one day— for they, too were tempted by the impossible dream of all mankind—they sprang out of a second-floor window, armed with an umbrella that was supposed to serve as a parachute.

In 1963 Papa won the Kennedy Prize. He left for the United States and returned with a sort of pyramid of transparent glass that had at the center an angel with wings, sculpted within the block. The statue was ensconced in the showcase in the living room, and Mama would proudly point it out to visitors.

Thanks to that glass statue, one Sunday Papa and Mama drove us in the car down a little road along which everything was turning green. There were a few houses there and a tall hedge with a little white latticed gate. Mama took me by the hand, and we walked across a large lawn, bordered by flowers. In the back there was a little white house with no one inside but a lady. It was "the country", where, once the transactions were settled, we would come every Saturday afternoon and stay until Sunday evening.

It was only much later that I understood the story of that magic statue. Papa had also received some money, half of which was to be applied to his research, while the other half was intended for him personally. That is how the white house became our playground during the weekends and school vacations.

Since it was an old village farm, the house had several little outbuildings. In the stable and the rabbit hutches, our parents founded "Kidsville", which became our domain. In another structure, Papa set up his office. The disorder that reigned there was heterogeneous yet organized. There was a workbench, some tools, an image of the Holy Shroud, lecture notes, pieces of inner tube, a few old pots, and some decade rosaries that were being crafted. Along the back wall was a display of white elephants that Papa stoically put up with. These are things that Mama wanted to keep but did not use any more and that she put there to make room in the house, which, after all, is not very big for the seven inhabitants who descend upon it. Most important, all over the office there were spiderwebs, which Papa carefully protected. It was strictly forbidden to clean his inner sanctum. Spiders eat insects, and Papa pretended that he was thereby protected from mosquitoes, flies, and other pests that would otherwise have come and tickled him while he was lost in thought. So, from time to time, he took his broom and swept the wood shavings, taking care not to disturb his guests.

That is where, on Sundays and summer days, he meditated, he wrote, and he reflected while answering the numerous and bizarre questions of his children and then, as time went on, of his grandchildren. Oh! Grandfather's office! You were allowed to touch everything, and Grandfather initiated you into the mysteries of electricity, repairing a "racing bike" that is going on forty years but that, thanks to do-it-yourself ingenuity, enthralled its temporary owner; he showed you how to make a bow and arrow, castles and fortresses, and millions of other props for imaginary adventures, all out of recycled objects. What marvels, fashioned from tin cans, carved wood scraps, and even out-of-commission household appliances! He would have won plenty of prizes in a Rube Goldberg competition!

One of his inventions used to send his children into gales of inextinguishable laughter. Everyone knows that the ground is low down, and my father, who loved gardening very much, suffered from a bad back. Therefore he devised a weed killer that was foolproof. He brought an old laundry iron out of retirement and attached it to the end of a handle in such a way that you could change its orientation without having to bend over. Burn the weeds, that was the trick. Unfortunately, it proved to be only relatively efficient, because the device lacked an independent power source; yet some time afterward we saw in a catalogue a tool designed according to the same specifications. We never knew who sold the plans!

One triumphant day he discovered some wild strawberry plants. He carefully divided them in two and replanted and watered them while a neighbor looked on sceptically. Several months later the plantings had put forth tremendous new growth but had not produced any fruit. They were weeds that happened to resemble strawberries.

There was also the era of the Jerusalem artichokes which, as everyone knows, grow like weeds. That, moreover, was their main point of interest, because once the novelty of tasting this unusual vegetable had passed (the flavor vaguely resembled that of conventional artichokes, and so it had become popular during the war), we grew tired of it. The Jerusalem artichokes have a reputation for being hardy, prolific plants, and they didn't let us down. They invaded one whole section of the field, and it still happens that we find a few, ten years later.

My father loved our stays in the country. Every Sunday he set out alone, with his walking stick, to take a stroll through the fields, behind the hill. He knew all the different kinds of animals that populated the countryside. He would stay there for a long time, marvelling at the splendid simplicity of nature, in love with the world that men lived on, and with its infinite diversity. That is where he often found the hook, the phrase that would give to his scientific theory all its force and clarity, or where suddenly he would understand one obscure link in the chain of life, inscribed in the DNA. He would return and write, or else dictate, thanks to a little dictaphone, the results of his thinking.

No doubt it was in the calm of his inner sanctum, where the spiders spun their webs in ceremonious silence, that he wrote these few lines as a foreword to his publications and studies:

> The genetic burden weighing upon our species can be measured only with difficulty. Each one of the genetic afflictions is, by itself, quite rare, but the catalogue of these illnesses is so extensive that at birth almost four infants out of a hundred are affected by one of them, to various degrees.
>
> To take only one symptom—the most dramatic one, no doubt, because only man can suffer from it, and the most inhuman as well, since it prevents the patient from being fully himself—a mental handicap strikes almost 3 percent of the population. And at least one-half of those so afflicted are suffering in their flesh and in their mind the consequences of a mutated gene or of a chromosomal aberration.
>
> To the study of the causes of this immense distress I have dedicated all of my research activity.

Reminiscence

In the beginning there is a message,
and this message is in life,
and this message is life.

It is said that a president of the French Republic once wrote, "The dead do not want us to weep for them; they want us to carry on for them." My father's inspiration was, rather, the eloquent and sublime final line of the Brahms *Requiem*: "Blessed are those who die in the Lord, for their works follow them."

Since his death, his family, his friends, his colleagues, and the parents of his patients have united to continue his scientific and ethical work. La Fondation Jérôme Lejeune, a foundation for research into mental handicaps, has been recognized as a public service. Through it, his battle against "the most painful sickness that there is, because it prevents the sick person from being fully himself," will be continued and, we hope, advanced by numerous researchers throughout the world. "To attempt to restore to each one that fullness of life that is called mental freedom: there is a real task for us, for our successors and for their successors." Through love for the sick person, respect for his life and his dignity, and compassion in the face of his suffering, his practice of medicine was at the service of mankind and uniquely for that purpose.

Now that his race has been run, the tremendous support of thousands of people, the expectation of so many families who suffer, and even the attacks of those who would like to see Papa dead and buried once and for all bear witness to the power, the relevance, and the truth of his message.

"Blessed are those who are persecuted on my account." As a child I pictured myself as a missionary, in a distant land full of dangers, where you had to defend your faith and your life. How exciting and stirring that must be. And we used to think of that "iron curtain", behind which there were men and women fighting for the freedom to practice their faith.

I had never imagined that persecution for one's beliefs could exist in France. And yet I was seeing it day after day. For Papa's life was the stuff of destiny. Here is a man who was the first to discover an illness caused by chromosomal aberration, trisomy 21, who opened the doors of modern genetics. Here is a man who was chosen by Georges Pompidou, President of the Republic of France, to participate in Les Sages, an assembly of scientists destined to counsel the political authorities; who was consulted by all the journalists whenever genetics became a current topic; who was invited to all the corridors of power and welcomed. Here is a man whose scientific talents were applauded, who thereafter experienced celebrity, who had a wonderful career ahead of him, full of honors, recognition, and power.

But here is a man who, because his convictions as a physician prohibited him from following the trends of the time, was banned from society, dropped by his friends, humiliated, crucified by the press, prevented from working for lack of funding. Here is the one who became, for certain people, a man to be beaten down; for others, a man not worth jeopardizing your reputation with; and for still others, an incompetent extremist.

Oh, sure! Everyone has the right to his own beliefs; you just don't have the right to speak them loud and clear. That, it seems, is a crime of high intolerance. If you do not think the same way as those who produce ready-made opinions, you are guilty.

How he had to love them, his patients, in order to explain them to everyone and to defend them against all who questioned their right to life and to give them hope of being cured some day or other! He himself knew very well that all life, even life that seems diminished in the eyes of the world, is worth the trouble of living. That treasures of love can be read in the eyes of an afflicted child when one dares to love him. That behind the illness, there is a child capable of love, of tenderness, and, why not, of a certain happiness.

He, more than anyone else, was aware of the suffering of the parents; he saw it every day. He had a boundless admiration for their devotion, their wealth of imagination and ingenuity in coming to the aid of their sick child. He knew also their distress in the face of an illness that medicine is powerless to treat, when confronted with the death of a child "not like the others", when encountering everyday difficulties in a world that is unable to accept those who are different. He also knew their worries about the present, about the future: "What will be the lot of our child in this world? What will become of him when we are gone?"

But that did not justify, in his opinion, taking into his own hands the right to life or the death of one of his fellow creatures, an innocent patient. Such a liberty he could not accept. In his view, every human being has the right to life and the right to be treated with dignity and, if he is sick, he has the right to assistance from society. Not "the right to die".

It was not because of his religious beliefs that he thought this way. He often said, "If, God forbid, the Church ever approved of abortion, then I would no longer be a Catholic."

It was his vocation as a physician and, above all, his scientific knowledge that guided him. He knew, and he had proved it many a time, that in the first cell, from the very first day, the genetic patrimony is written in its entirety. Even before the mother knows that she is pregnant, the child is taking shape as a result of this unique genetic heritage, which belongs to him alone and will never belong to anyone else. "The little one [offspring] of man is a little man", he liked to say. That individual, however little he may be, will have blue eyes and brown hair; he will have a choleric, passionate temperament, and will be gifted in mathematics. That is what one could predict if one knew how to decode the entire genetic inheritance. Because every human being is unique, because he has an identity from the first day of his existence, because he is a member of our species, his life must be respected. The true physician does not have a choice.

When all is said and done, even in the sight of his detractors his involvement was a proof of his nobility, courage, and conviction. But he ought to have kept quiet. To speak with such talent, to defend the weak by means of the truth, that's what was unacceptable. Since he couldn't be intimidated, he had to be marginalized, made into a dangerous fundamentalist who had blundered into an illusory battle that was lost from the start. He had accepted the consequences of this, and he died without betraying his conscience.

This destiny, this life cut in half, was written for us, his children, to see on a day-to-day basis. When we were children, our father was an honorable man, a congenial intellectual who was all the rage among the elite. When we were adolescents, he was avoided like the plague. He had committed the misdemeanor of having the wrong opinions.

But even though we shared in his day-to-day life, that changed nothing of the joy of family life, of his zeal for his patients, of his aggressive campaign to cure them. Now that I reread with the eyes of an adult the milestone events of his life, I realize that he must have suffered much. He experienced the renunciation of the things of this world: glory, celebrity, professional recognition. He experienced the betrayal of friends, administrative harassment, the condemnation that the modern press can wield. Today people no longer remember the violence and the passion of the debates of that era.

If he suffered, he never let us see it. In the face of insults he used to smile, saying, "It is not for myself that I'm fighting, and so these attacks don't matter."

All the while that he was deluged with opprobrium, we were proud of him. His courage, his patience, his complete lack of vengeance were for us a tremendous education in life. He knew too well what was in the heart of man not to pardon its weaknesses.

Some time after his death I had a strange dream. My husband and I were attending the opening of a museum of contemporary photography, dedicated to the personages who had shaped the twentieth century. General de Gaulle, Clemenceau, André Malraux, and quite a few others—each had a gallery where historic photographs, many of them unpublished, were on display. Two government ministers and various notables were present. Those in the crowd jostled one another to follow them, to walk in the wake of the cameras. We went off a little to one side, and, on the right, we saw a little room, which they were just finishing. It was dedicated to Bernanos.

A man was there, leaning over his workbench, completing a collage. When he saw us approaching, he raised his head. He was bearded, with a timid smile and dreamy eyes brimming with the task that he was trying to accomplish. He straightened up and greeted us courteously; we spoke of our shared enthusiasm for the novels of Bernanos. I looked distractedly at the disorderly workbench, and suddenly, at one corner, beside a pile of all sorts of documents, I saw a little newspaper clipping. It was a notice of my father's death. Surprised, I turned to our interlocutor and told him who I was. He then told us how much he esteemed him, even though he didn't know him. I spoke to him about his death and about his last words dedicated to his little patients, whom he imagined he was abandoning: "I was the one who was supposed to cure them, and I'm going without having found it."

At that moment, a sweet, powerful breeze enveloped me; I felt the arms of Papa surround me with tenderness and warmth. His voice, coming from behind, entrusted his patients to us; he mentioned two or three by their first names. Hervé and I answered him, but he did not hear us. Then the breeze was gone.

An immense joy, but also a dull uneasiness, rushed in upon me. It was he; I was sure of it. I felt him, I could almost say, in the flesh. But to take up the torch seemed a formidable task. It is an immense and difficult job; I am not a scientist, or a physician, or a biologist.

In the morning these mixed feelings continued, and they lasted for several days. I know that he came to speak to me that night. There is nothing scientific to this certitude; didn't I just dream it? But the breeze was not a dream; it came from somewhere else.

Every day we discover new aspects of his activity and his influence. His life's work is beyond us, but it is beyond him also. He was only one man, with his faults and his good qualities, who tried to blaze a trail. This story is the one that we are in the act of writing today, in trying to continue his work. We will have disappointments, failures, and injuries. But there will also be victories won against sickness, against lies, and against despair.

That story, too, will have to be told some day. But let childhood return and tell, through its eyes, how my father made his way through all those years, past so many snares along those paths that he had not really chosen.

Brother Jérôme

One sentence, one alone, will
determine our course of action; the
reasoning which cannot deceive
and furthermore which judges
all things, the very words of
Jesus: "As you did it to one of
the least of these my brethren,
you did it to me." (Mt 25:40)

Brother Jérôme: that is what Pope John Paul II called him in his message of April 4, 1994, the day after his death. He wrote, "One must speak here of a charism, because Professor Lejeune was always able to employ his profound knowledge of life and of its secrets for the true good of man and of humanity, and only for that purpose."

His bond of friendship with the Holy Father was profound. In this intellectual there was a humility before God that was simply amazing. From his appointment in 1974 on, he was a member of the Pontifical Academy of Sciences, which gathers together the most brilliant scientists from all over the world. The main fields of specialization are represented, and more than 40 percent of its members are Nobel Prize winners. Many of the scientists are not of the Catholic faith. The role of this Academy is to enlighten the Holy Father on the most recent scientific findings.

He was immensely interested in this office, and even though he set little value on the titles with which he had been honored, he was profoundly happy and proud to be a member of this prestigious assembly. When asked what titles he held, this was the one that he always mentioned first. He was glad to be of service to the Church and was also a very active member of the Pontifical Board of Health, directed by Cardinal Angelini.

Pope Paul VI liked him very much, but it is especially with John Paul II that the ties of friendship, respect, and profound communion were woven in the course of the years.

He did not pride himself on this bond. He considered it an undeserved favor that he was invited to eat at the Pope's table and, even more important, to participate at his private Mass in the morning whenever he was at Rome with my mother. They would always return immensely happy about these exceptional moments spent with the Holy Father.

On May 13, 1981, my parents had lunch with the Pope. At about three in the afternoon they said goodbye. Papa and Mama were escorted to the airport to take a plane. The Pope left in his automobile to greet the crowd. In a few moments he would have a brush with death.

My parents took their plane. In the taxi that drove them home from Roissy they heard the news, which upset them terribly. That night Papa came down with horrible pains in the stomach, and Mama had an ambulance take him to the hospital Hôtel-Dieu. For two days he thought he was going crazy. No one understood what was wrong, and he experienced the pain of the Pope's wound. That was madness for a mind as rational as his. Finally it turned out to be three little stones that, as a result of the emotional shock, had blocked the bile duct. In layman's terms, he was suffering from gallstones.

During the two days when he was contorted with pain, we really thought that he was going to die. Even before then I had marveled at Mama's extraordinary calm in difficult moments.

He would have surgery, like the Holy Father; their temperature curves would be similar; they would leave the hospital on the same day. The coincidences amused us, but he did not like people to refer to them in his hearing. His scientific, rational mind was always opposed to sensationalism and miracle hunting.

In 1994 the Pope, knowing that my father was terminally ill, would nevertheless appoint him the first president of the Pontifical Academy of Life. He would hold the office for thirty-three days, and he said, "The Pope has made an act of hope by appointing a dying man." A few days before his death, speaking about the Academy, he added, "I'm dying while on special duty."

Throughout the course of his illness, the Pope would ask for news about him. He sent a telegram, the day before Papa's death, for the Easter Vigil. One of the favorite saints of the Pope is Sister Faustina, a Polish nun from the early twentieth century who had visions and left profoundly moving writings on the Divine Mercy. For that reason, Cardinal Angelini and his collaborators conveyed to Papa a relic of Sister Faustina and recommended, "Pray to her every day."

And so our daughter, born thirteen days after the death of Papa, had the good fortune to be named Faustine, "felicity, joy". Her birth was announced to Mama while she was substituting for Papa at a congress in Warsaw. Faustine and her grandmother then received an ovation from the participants. A coincidence?

Papa had had a pious, well-behaved childhood. He had a child's simple, confident love for Jesus that was surprising in a mind so profound and intelligent.

He liked to tell of his meeting with an old rural parish priest, who had had the words "*Vérité, humilité, paternité*" inscribed over the door of his rectory.

And whenever someone asked why, the priest would reply, "Well, it's all very simple, because the truth [*vérité*] will set you free [*liberté*], humility will make you equal [*égalité*], and paternity will teach you that you are all brothers [*fraternité*], since you all have the same Father."

It was a very Christian reading of the French national motto. But, after all, once you pay your respects to the saying, each one is free to find in it whatever personal interpretations he finds inspiring. It was quite similar to the reading of life that Papa was able to make.

Papa was a great Christian. He lived his faith, which was rooted in his flesh, and from it he drew courage, kindness, attentiveness to others, and, above all, what was most striking: the absence of fear.

He was not afraid. What can someone do against a man who doesn't want anything for himself? He accepted the joys of life as a blessing from heaven. He endured the pains without seeming to attach any importance to them, and still he had his share of them.

He was discreet about his faith, but one sensed that it dwelt within him. Though he was a man very busy with his profession, he accompanied us on our pilgrimages, in the rain, with his gray plastic cyclist's poncho to protect him.

His researcher's mind could be in ecstasy at any moment over the ingeniousness and complexity of the universe. All of that was for him a confirmation of his fundamental inspiration: "Everything comes from God."

During the sixties and seventies he had "crossed swords" in the scientific and intellectual arena with those who followed the fashion of Darwinism. Notably, he had done a lot of work replying to the book by Jacques Monod, the reference work of the day, that was entitled *Le Hasard et la Nécessité* [English edition: *Chance and Necessity*]. It was about chance meetings and the necessity of survival, which made possible the appearance of new species, scattered here and there all over the world. The theory had some success. It also had the merit of destroying the hypothesis of the single pair and of demonstrating yet again that Christianity taught stupid nonsense and prevented humanity from advancing toward knowledge.

Now, in Papa's opinion, all of the genetic findings that he knew about gave him the intuition, though not the proof, that humanity had probably descended from a single couple. For him, considering it from a strictly scientific perspective, the most reasonable hypothesis about the dawn of humanity was that of Adam and Eve. He had demonstrated it many times. Today, it is the theory that the majority of scientists hold.

He was accused of putting science at the service of his faith at any cost, but that is not what he had sought to do. I believe I can say that he never experienced the spiritual throes of the intellectual who discovers that science proves something contrary to his faith. On the contrary, he was fascinated by the great wisdom of the biblical texts, which, using metaphors, traced what is probably the most plausible sequence of events by which our universe may have been created.

I cannot pass over in silence an article, written for the weekly newspaper *Charlie Hebdo* on the occasion of his death. The journalist, André Lancaney, considered him "an enemy of the worst kind" and nevertheless recognized his scientific talent. On the subject of Darwinism, he wrote:

> Lejeune defied neo-Darwinian biology, which was in decline, as well as the molecular biology of Jacques Monod, in proposing an "Adamist" theory of evolution, in which he claimed to discover the proofs of the biblical book of Genesis . . . in the structure of chimpanzee chromosomes! The theological justification was dubious, to say the least. But with extraordinary wit and verve, Lejeune managed to raise the most serious objection yet to the theory of the gradual, progressive evolution of species and to show that evolution must have made hops, skips, and jumps, contrary to the doctrine of Lamarck and Darwin. It was

only ten years later that the very Marxist paleontologist
Gould and his colleague Eldrege adduced proof of this fact
from the fossil record, without realizing that the studies
of Lejeune and of his student Bernard Dutrillaux made
an explanation of it possible.

Nevertheless, no one stood up to acknowledge the scientific truth that he had unveiled. Inasmuch as he was a
Catholic, and his theory was not contrary to the faith, he
was not to be believed.

We were brought up in the Catholic faith. Unostenta-
tiously and dispassionately. After dinner, every day, Papa
used to bring us together for evening prayer. We would
recite an Our Father, a Hail Mary, and then say a short
prayer for those in our family who had died, and finish
with intercessions for all the sick. It was an occasion for
heart-to-heart conversations with him, one of those privi-
leged moments because it recurred every evening and gave
a rhythm to the seasons. It was also the time of mercy,
when it was easier for us to admit some foolishness that
we were carrying like a heavy burden in the secret of our
hearts.

On Sunday we would go to Mass as a family, and Papa
sang in the choir with some friends. We would meet him
along with others when Mass was over for a snack at our
house. These family traditions, which seemed immutable
to us, were part of our spiritual décor.

So the example that he gave us was for us an everyday
catechism lesson much more effective than any words or
instructions. Bishop Guérin, in his homily during his fu-

neral at the Cathedral of Notre Dame, would say, "You must have experienced, many times, gardens of olives where the joy of saving is mixed with the heavy duty that must be paid in order to pass on a little joy, a little hope, a little truth, a little love."

It was always Papa's intention to give us our freedom, and this kept him from any attempts to influence us. He fed us and lived up to all our expectations in order to give us what he considered to be the most precious gift that a father can give to his children: the gift of knowing that they are loved, infinitely loved by the God of the living. And because he believed that "the truth will set you free", he also gave us the instruments of this freedom, which he made his own, while leaving to us the task of making it fruitful in our own way.

It's funny, but in order to evoke the Christian life of my father, the only means that I have left is silence. The respectful silence before an interior life that governed his exterior acts, but which belonged only to God and to him. We have seen his witness; the rest is inexpressible.

This intimacy with the Lord. He left us evidence of it. In a letter to his brother, which he never sent, written after a trip to the Holy Land:

> In the Holy Land, at Tiberias
> And in this little chapel decorated in bad taste, on this recent flagstone pavement that was perhaps not yet thirty years old, I stretched out full length to kiss the imaginary footprints of Him who was present there. This naïve, I could almost say instinctive gesture seemed to me ridiculous in itself, though the sentiment that impelled me cer-

tainly was not. To speak of one's love is always impossible, and the classic declaration of love while placing one knee on the ground is probably much more true than the theatrical image would lead you to think.

At any rate, like a monk arriving late for his chapter meeting, I kissed the pavement as a sign of loving respect, because I did not know what to say and could think of nothing else to do.

You must not imagine, my dear brother, that I had at that moment a vision, that I was spiritually transported, seized by an ineffable ecstasy. All that is within me was as reasonable as usual, judging that my action was foolish, and yet my soul resounded with an unknown yet familiar vibrancy in attempting to join, by adoration, a unison that I myself could never attain.

A son finding again a much beloved father, a father finally known, a revered master, the discovery of a very sacred heart . . . all of that was in it, and much more. How should I put it: there was tenderness, sweetness, affection, a love that was timid and yet confident. A need to acknowledge how my heart was touched by so much gentleness and goodness on His part, that He should be willing to be there, that He should accept the fact that I recognized Him there, that He should welcome me so simply and fraternally.

How can I say it . . . there was tender love, unmistakably.

When I got up—my adoration had lasted no longer than a gesture—I laughed a little at myself, recognizing that, with all my scientific knowledge, I had not been able to offer anything better than an ancestral prostration, and I told myself that, like the soldier standing at attention and discovering pride in paying respect, I had discovered in the posture of a suppliant slave the tender affection of a willing devotee.

Outside, the sun was shining, as bright and cheerful as when I went in. The dear Franciscan friar, weary with age and the heat of the day, had retreated into the shadows of his tiny convent. And I set off again toward the Sea of Galilee, carrying with me forever the certainty that Jesus has prepared encounters and a marvelous intimacy for mankind, here or there, down here or up there or down there or up here, very far away, very far but perhaps very soon, on this real obverse of everything that exists, which is only discovered, at last, when we can see it from the other side of time.

Singing a Song as He Goes Along

Faith tells us to respect
the image of God,
Hope helps us to defend it,
Charity judges all.

It has arrived, this painful moment when I must call to mind his final departure. It would take a long time before the words would come. The death of a father is bidding farewell to the child who is still living within us and who rebels against the passage of time. It is a dull grief, nameless, odorless, without memory; it means that you will never be the same again.

It is a certain sweetness that is fading away, like the beating of a heart that gradually grows faint in the night. The insupportable weakness of a life that is withering beside the little fire of pain. It is the dreadful, desolate glance of those who are there, in the silence, without understanding that love is made of eternity.

It takes time for a drawing scratched too deeply on the stone walls of a former school building to be effaced, and so too this grief of a child, who knows neither how nor why. One can become an orphan at any age, even though it is even more unfair when one is young. Farewell to the sunshine of a childhood forged in the enthralled glances of parents united in a plainchant of love. Farewell to the deserted beaches of words that you never said and that you would like to shout to all the world today, so that the one who has passed to the other shore of time might hear them.

Why should we call to be our witnesses strangers who will never know with what love he loved us? Why confess to them that we will never really know it ourselves, since love is so mysterious, obscure, and painful, even in its own dominion? As a memento? For memory's sake? For our children?

Maybe to some extent for all those reasons, but above all out of a passion for living. Let us not waste a single instant of this short life, which brings us so much and which we treat so casually. It is that, too: the echo of a life that is dying.

In September 1993, Papa came to Savoy to give a series of lectures on "the mechanisms of intelligence". He stayed with us for two days and began to suffer from the high altitude. Especially at night he had difficulty breathing. After returning, he felt out of breath for several weeks. He thought it was a heart problem and underwent tests. His doctor prescribed medications to regularize the heartbeat.

During the month of November he came down with a heavy, slow, stifling cough. It was possibly a side effect of his medication, which he stopped immediately at the advice of the cardiologist. But the cough did not stop, and already he guessed the sickness that was consuming him.

He would say nothing about it as long as he was not sure of it, but the X rays and the tests confirmed the worst: lung cancer, in a stage so advanced as to be inoperable.

With his colleagues and friends, Professor Chrétien and Professor Israël, he discussed candidly the possibility of a cure and what kind of treatment could be applied: a 50 percent chance of success with heavy chemotherapy, a very heavy dose, followed by radiation treatments. He would have to have, at a minimum, six months of intensive, painful treatment, if he survived.

His voice was changed, but with his indescribable smile he told us the news. He had a presentiment: "Don't worry until Easter. In any case I'll live until then. After that, we will know the definitive diagnosis." And then, because his gentle sense of humor never forsook him, he added, "I am being cared for by Chrétien and Israël, the whole Bible of medical science. I'm in good hands."

To Karin, who was crying, he said simply, "I will be a good patient, and, believe me, I will fight to the end."

He underwent his first six-day chemotherapy treatment at the beginning of December. We celebrated Christmas as a family in Paris. He was too weak to go to the country, where we would usually gather. There were forty of us, with many children in the group, but by organizing things harmoniously we were able to celebrate without exhausting Papa. He was beginning to lose his hair and noted that his nails were marking the calendar of his chemotherapy treatments by visible signs that they had temporarily stopped growing.

The treatments followed each other at Cochin Hospital. They put him in a little room. The corridors were noisy and echoing, and he was not able to sleep. Next door, an old gentleman had the television blaring with the door open.

"Papa, you can't get any rest with all this racket. Would you like me to go ask him to close his door and to lower the volume?"

"No, my dear. The poor man, he is deaf, and if he leaves the door open, it means that he must feel lonely. Let him be . . . him, too; he is fighting the only way he can."

His suffering was intolerable at times, but he was always considerate of others; he put himself in their place.

On his bed of pain he continued his work, his research untiringly. Exhausted, he would answer the telephone, between two bouts of vomiting, to discuss a therapeutic hypothesis with a colleague. He was an arm's length away from heaven, and he had to attend feebly to this cancer that was robbing him of time, the time that was so precious, that he would have required to find a remedy for this illness that haunted him, the kind of mental handicap that is inscribed in an imperfect genetic patrimony, which man must learn to repair.

When he told Pierre Chaunu, his great friend, that he was sick and had to give up his office as president of the Academy of Ethical and Political Sciences, to which he had been elected, he sensed the consternation of his friend. As Pierre Chaunu tells it, "He said to me, 'Please excuse me, I have the impression that I'm causing you some pain.' Some pain, sure! Yesterday, today, tomorrow . . . for a long time."

His friend Pierre would call him every day, to get the news, to maintain such a precious tie of friendship when life was hanging only by a thread, when it became a constant battle to keep it from stopping.

Between two treatments he would return home. Mama arranged for him a sickroom-office, where he could go from his bed to his desk without too much trouble. At every moment of respite he would write, take notes, read, and work. He happily welcomed friends who came by. He would hear his grandchildren playing and crying downstairs, and then they would come, in silence, to kiss their grandfather, who was so sick. Since he could not see them and play with them, he was happy to hear them enlivening this house that was made for life.

He used to say to us, "It's crazy how much time it takes to be sick!"

He was physically very weak. Besides his treatments, there were a multitude of little miseries that vexed his frail constitution: a phlebitis, a catheter that had been inserted incorrectly, an enormous hematoma on his neck as a result of procedures clumsily performed by an attendant, and so many other sufferings that were added to the pains of the treatments. He never complained. He joked and reassured us.

Sometimes he would emerge to have a meal downstairs, in the dining room. But from the month of November until his death he would go outdoors on only one occasion. He didn't have the strength for it any more.

Afterward he would say, "I knew that I was foolish to do it."

At the beginning of March he put on his elegant navy blue suit again, the one that he was already wearing at the time of his inaugural lecture in 1962. He was the shadow of his former self, but for us who saw him every day, it was a sort of resurrection. He went to the Academy of Medicine to support the candidacy of his most faithful collaborator of all time, Marie-Odile Réthoré. She would be elected one year later. For the first time since Marie Curie, a woman would be admitted to this prestigious body. The press would not even report the event. Strange, isn't it?

Then came the debate about bioethics. A long time before he had asked for an interview, which would not be granted to him, with Edouard Balladur, then the Prime Minister.

He feared, as did many others, that technological advances had made genetic manipulations on living human embryos possible. He hoped that the embryo would be distinguished from other human tissues by means of a regulation protecting it from manipulative sorcerer's apprentices. Immediately there was a hue and cry: "Watch out, they want to debate all over again the 1975 law on in vitro fertilization." But wasn't the point of that law in the first place to declare in article 1 that "the life of each human being must be protected from conception until death"?! They failed to comprehend that technology since then had advanced much further and that, if precautions were not taken, eugenics and the production of men made to order would be possible tomorrow.

He wrote to one of his friends, "Until now I have tried to be the soldier to whom the centurion says Go, and he goes.

"For the moment I can go neither far nor fast. Just when it is imperative to defend the embryos who will be attacked on the day of the Holy Innocents, I'm out of breath. For the moment, faithful to the Roman legionary's motto, '*Et si fellitur de genu pugnat*', I write, 'And if he should fall, he fights on his knees.'"

He died on April 3. The day after his death, a full-page advertisement appeared in *Le Monde*, a petition signed by three thousand physicians within a few days, demanding recognition for the embryo as a member of our species, not to be exploited for manipulations of any sort whatsoever.

That would be his final battle for the dignity of and respect owed to every person, whatever his size, age, race, or religion.

In March the first radiation treatments began. The lung put up no resistance; the pleura opened and filled with liquid. He was taken by ambulance to a clinic in the west of Paris, where he would undergo surgery. For six days he was unable to eat, and it occurred to no one that he would have to be fed intravenously until Mama realized that not one of the many cumbersome tubes that he was on was providing him with nourishment. At the doors

Life Is a Blessing

of death, he was being starved because of thoughtlessness and negligence.

And the doctor told him, while he was suffering intensely, "The treatment that they had you on was foolish; they should have used an entirely different procedure."

How could one be so cruel toward a man who was battling cancer with his last ounce of strength in this race against the clock and against death.

Wednesday afternoon, Anouk and I go to see him. He has a tube, in which the blood and the water that drain from the lung are mixed. He is exhausted, his tongue is one large sore, but he is smiling, happy to see us.

Anouk suggests that he receive the Sacrament of the Sick. He agrees. There are moments when he drifts or falls silent from the exhaustion, but he has his wits about him.

That evening he becomes delirious. He has a fever of 40° C (104° F) and has not been fed for six days. Mama and Marie-Odile Réthoré decide to have him transferred to the clinic *des Peupliers* where he has been treated before. The next day he is doing better; the fever has broken. He has within reach an oxygen mask so that he can catch his breath again.

He received the Sacrament of the Sick on Friday. The priest would tell us afterward, "Without breaking the seal of the confessional, I can tell you something that Jérôme said: 'You know, Father, I have never betrayed my faith.'"

I have always been impressed by the fact that Papa, as a physician, knew better than anyone the fate that awaited him. He accepted physical suffering with an unfailing courage. But what astonishes me even more today is that never once did he appear to be afraid of death. That was just at the end of the road. He would never be cured of this sickness; he was simply trying to gain some time. But even in that he was not afraid.

In that large, sinister room in the intensive care unit, another patient was dying. He snored appallingly and cried out during the night. Papa would comfort him. One night he fell out of bed, and Papa, unable to get up and help him, called the nurse on duty. In his misery he had found someone worse off than he to help and to console.

Friday evening, my brother, Damien; my husband Hervé; and I went to see him. We had to have him sign some papers concerning his laboratory. In the car I said to Damien, who was a deacon, "It's up to you to ask him about his last wishes. Mama shouldn't have do that; she is there to help him hang on to life."

And we would then have those unforgettable moments when, whatever might happen afterward, all was said, all was finished. He spoke in short gasps, stopping from time to time to breathe some oxygen. I was holding his hand. It was hot and soft like a baby's hand because of the chemotherapy.

I asked him if he wanted to bequeath something to his little patients. He answered, "No, I don't mean to neglect them, but, you see, I don't own very much. Besides, I gave them my whole life, and my life was all I had."

Then he continued, "What will become of them now? What will they think?"

"But, Papa, they know very well that you are sick. You

haven't seen anyone at your office for many weeks now. They, surely, understand these things better than we do."

"No, they don't understand any more than we do, just with greater depth."

"Papa, it appears that the hand of God is resting on you. Look: you had lunch with the Pope on the day of the assassination attempt, in 1981. You were at the hospital and had surgery at the same time he did, and today you're suffering the Passion, this Good Friday."

"My children, if I can leave you with just one message, the most important of all, it is this: we are in the hand of God. I have had proof of this on several occasions in the course of my life. The details are not important."

And he went on, "It is true that death is not far off, maybe in two weeks, a month, three months. I don't dare hope for a year. But, my children, don't forget that Sunday is Easter. Something could happen on Easter Day."

Was he hoping to be cured, or did he know that he had a rendezvous with his Lord in glory? We will never know, but several times, in the course of his illness, he had considered Easter as a crucial day either for life or for death.

Damien asked him about his wishes concerning the Mass and the cemetery. How hard it is to speak of those things with someone you wish would live, live without end!

"You can just do as you please, my children. That will be fine. I have only one request: that my little patients should be allowed to come if they wish, without being intimidated, and that some places be reserved for them."

"And Mama . . . don't worry about her, we'll take care of her."

"I'm not worried about her; I know that she's an extraordinary woman, and I know very well that she will prove to be even more so after I am gone."

How could a man who loved his wife so tenderly leave her without rebelling? Oh, there was grief, an immense grief that welled up in his eyes; but anger, feelings of being wronged, denial? Not for a moment. He had that gratitude that comes when a life is accomplished, the peace of a just man, the certainty that he was not leaving us completely. We have touched on that ineffable love that dwelt within him. We are happy. We let him rest.

On Saturday he had a visit from his brothers and cousins. That evening, his breathing became more and more difficult, and Mama wanted to stay beside him through the night. The room where they had put him was large, sinister, and impersonal. How sorrowful the nights must be in this soulless universe, fighting without strength far from those you love.

He refused the offer with his last bit of energy: "If someone comes, I'll be furious." Mama left then with Anouk, with an uneasy, heavy heart.

We had just moved to new quarters, and I was eight and a half months pregnant. We were staying in the *rue Galande* with Mama, since they were still working on our new apartment. The children were in Savoy, on vacation from school.

At four in the morning he begins his agony. The doctor who is attending him wants to notify us. He refuses. He does not want us to retain an image of him in the intolerable pain of this final suffocation. It is his last act of charity. With his last gasp, before his life is extinguished, he says to the doctor, "You see, I did right."

At seven in the morning we are awakened by the telephone. It is Easter morning; the first bells of the Resurrection are ringing in the brilliant morning sunlight. The trees are in bloom, the birds are chirping, and we leave to see Papa for the last time under a crystal clear sky. When we arrive in that immense, cold room, he is there, still warm, his face marked by the suffering and the sickness. The priest who was supposed to bring him Easter Communion arrives, and Mama is the one to receive, kneeling at his side, the Host that was destined for him.

We have to go through the usual formalities to bring his body back home. We are filled with a strange, powerful peace. The mortuary chapel is set up in his room. Friends, notified by radio or by the family, begin to file past. Mama, smiling, welcomes and consoles them.

But it has become impossible for us to hold a wake over him. That mask of wax there, that's not him. He has gone; his soul is already over there, and this body resting to one side no longer resembles him. He has carried off with him his luminous glance and that smile at the corners of his mouth. "Let the dead bury their dead."

Jean Foyer calls us. "What's happened? I saw the Pope on television this morning. He was sad. You could see it on his face. I thought that something must have happened to your father." In fact, the Pope, who had sent an affectionate telegram to Papa the day before, had learned, an hour after his death, that Papa had taken the road to eternity. "I am the resurrection and the life; he who believes in me, though he die, yet shall he live" (Jn 11:25).

Numerous political personages and scientific notables call on us and come to recollect themselves side by side with our faithful friends.

In the choir of the Cathedral of Notre Dame, before a compact, recollected crowd, Bruno, who has trisomy 21, walks forward. It was on the basis of his karyotype and those of six others afflicted with trisomy that Papa had made his discovery. Bruno always considered that a great honor. During the prayer of the faithful, he takes the microphone, to the surprise of most people present. He is holding in his hand the picture that we distributed on the chairs at the beginning of the ceremony.

In a strong, clear voice he says, *"Merci, mon professeur*, for what you did for my father and my mother. Because of you, I am proud of myself."

That was Papa's life. Directing his glance toward someone who is unseen and taking him by the hand so as to bring him into the light.

The splendor of the ceremony, the multitude of expressions of affection and of sympathy, the support of all our friends, and also the will to continue his work, all of that was real. He would have been astounded that his death aroused so much emotion.

He would have liked to leave this world with his characteristic discretion and humility. But now what shines is the message of which he was the bearer. Unless the grain of wheat dies, it cannot bear fruit.

Some time after his death, while putting his office in order, Mama found this poem that was intended for us:

> Ainsi chantonne doucement
> Un vieux cœur en s'en allant.
> Femme aimée, très chers enfants,
> Beaux-enfants, petits-enfants,
> Frères, nièces, tous parents,
> Prenez en gré je vous en prie
> Si je faufile hors de temps
> De l'autre côté de la vie
> Où le Seigneur nous attend.
> A tous grand merci disant
> A Dieu, sa merci demandant.
>
> And so he softly sings a song,
> An old heart, as he goes along.
> Beloved wife, dearest children,
> Sons-in-law and young grandchildren,
> Brothers, nieces, family all:
> Please take it well, I ask of you,
> If when my time on earth is through
> I'm sneaking to meet life's other side
> And with our waiting Lord abide.
> To all a hearty thank you saying,
> And so to God for mercy praying.

That is his song as he goes along.

Epilogue

When I wrote these lines, I did not know that thirty-seven years earlier my father had experienced the same grief— a sorrow beyond all measure but without bitterness. His father had died of the same illness as he did, and he had been the only one forewarned about it. Some time ago my sister found a journal that my father had written from 1959 to 1983. We did not know that it existed. It testifies to the tenderness with which he loved his parents and gives a day-by-day account of the final weeks of his father, who was nicknamed BonPa.

Because he knew what kind of an end was in store for him, he kept us at his side, and then right before the final suffocation—an atrocious, unbearable ordeal—he wanted to be alone. No doubt he thought that that particular suffering was not necessary. He did not want to inflict it upon us. His last act of charity was to leave us with an image of himself that would not be too frightening, so that it would not haunt our memories. He had experienced that. He tells about it in these few lines written at the age of thirty-three, one year after the death of BonPa:

Sunday, January 11, 1959.
Today is the first anniversary of BonPa's death. Already one year since my father, my Papa whom I loved so much, left us and since I, helpless as I was, tried to help him as he died since I could not save him!
My God! One year, which has been like years and years,

considering the change that has been imprinted on my heart, but which appears to memory to be not even a day. Time does not exist in the face of ever-faithful death! Yesterday or tomorrow, it is the same irreparable, unceasing absence!

For a long time Birthe has wanted me to follow in Papa's footsteps and to write, in turn, a journal like his. But I could not resolve to do it. Papa's yearbook stopped at the beginning of January 1958, and I didn't have the heart to finish a year that he had not lived through. Now, at the end of a year, I feel that it is my duty to become a man like he was and to take up again the task that he had continued until his end. . . .

It was a little before January 1, 1958, that Papa fell sick once more. Since returning from my first voyage to America I had seen him grow weaker and his features change and, in spite of the animal embryo extracts and the vitamins, watched age and sickness go about their work pitilessly. On New Year's Day, in the afternoon, Papa and I were talking together, softly and affectionately, in the old dining room in Etampes, which was rendered even more somber by the dark chestnut-stained woodwork. While we were talking in this way, I suddenly noticed that his fingernails were starting to bulge, in the form of a watch glass. I could scarcely believe my eyes; I didn't dare, didn't want to believe it, and, stealthily, without him noticing it, my glance kept returning to his nails, which were changing their shape.

This digital hippocratism, I knew what it meant, I had understood, and I was floored. My Papa, my father was condemned to death. His lungs were not respiring sufficiently, and so it happened that the underoxygenated blood raised the ungual matrices. His heart was already weak and would not be able to hold out for long!

At that moment I would have given anything at all not

to know, not to have seen those poor fingernails that the asphyxial blood was slowly causing to bulge. Oh, it was still very discreet, no one would notice it except me, but I could not deny the evidence: instead of the flat shape that I knew so well, this terrible terrible convex curve of the fingertips, which meant death!

Papa told me that he was really at the end of his strength, and that we, the physicians, did not see what was wrong with him, that everything made him short of breath, that he felt life departing from him. And I said no, and no again, while laughing affectionately and teasing him gently, as was our custom when we two were all alone and he spoke about some little ailment. And I had a lump in my throat that hurt. . . .

On Saturday the fourth we had an appointment at Trousseau, and Papa came to sleep at our house on Friday evening. We had a truly wonderful evening, a short one to avoid fatiguing him excessively, but full of love. It was the last happy evening of his life. After dinner I played guitar for him a bit, very poorly, then we made plans to go to a seaside resort that Yvonne was just telling us about.

Saturday morning we left early for Trousseau, and before putting on my laboratory coat, I pointed out to him the path that he should follow to get to the consultation room, adding that I would rejoin him in a few minutes. When I came down a short time later to meet him, I saw him heading slowly, oh, so slowly, shoulders hunched, for the building that I had pointed out to him. I had gone up one flight, changed clothes, given instructions to the laboratory assistant, and during that time he had scarcely gone twenty paces! It was heart rending, more so than anything that I had feared, so unbearable it is to see those that you love suffering.

I caught up with him, and we walked slowly up the hill; I hurried him a little so that he would not catch cold, but that was only a façade so as not to alarm him, and I slowed my pace to match his.

Once we arrived at the nursery, I took the elevator with him to go up one floor.

Never will I forget his gait that day. How many times I have watched him make his way down a street in Paris or Montrouge when, after picking him up at his office and then saying goodbye to him, I waited for a few moments before driving off. But this was the last time that I saw him walk away from me, and that picture is now engraved on my heart.

Twenty paces! He had only gone twenty paces! Since he didn't know that I was watching him, he hadn't "forced it" that time, he who not long ago was so proud of his stride and who passed many young people along the street. Twenty paces, that's how far we had gone.

With the laboratory assistants during the tests, and with Dr. Lafourcade, whom he liked very much and who spoke to him with such charming deference, he was like a new man again. He became more lively, spoke easily, and seemed to be completely well, though a little out of breath.

Lafourcade was reassuring, and he sincerely meant the encouraging things that he said to us both. To see Papa at that moment, even though the X ray was all blotchy and at the bottom of his right lung there was a little pocket that was making crepitant sounds, one would think of nothing more than emphysema in an aged, sclerotic lung. I, on the other hand, was thinking of a lung tumor, but maybe I was wrong? . . .

On Monday the sixth, the results from the laboratory were reassuring; blood chemistry was just about normal, a slightly elevated rate of sedimentation, but all in all nothing serious. Lafourcade, whom I consulted to talk about them, concluded that I was wrong to be alarmed, and I ended up believing him, since I was just hoping that I was mistaken and every argument was from the start a consolation to me. . . .

At Etampes, things were not going as well as I thought. . .

In the dining room near the stove, Philippe and Mama explained to me how things were developing; they thought that Papa was "letting himself go", since he would recover his spirits so visibly whenever strangers arrived.

Without too much hesitation I went upstairs alone to his room to examine him and, above all, to embrace him.

He was dozing on his large bed, quite out of breath and obliged to stay in a half-sitting position. When he saw me, he got up, hugged me affectionately, and said to me, "How good you are to have come!" My heart sank at these words, and I replied, "*Mais non.* No, not at all, I am not good; I just wanted to see how you were doing."

The little pocket on the right was now enormous and took up the entire lower part of the lung—everywhere congestive rasping showed obstructions in both lungs. I hardly had the courage to keep the stethoscope in my ears. The blood pressure was good, the heartbeat slightly palpitating but regular. Would the antibiotics work again?

We both spoke in a low voice, and I joked a bit to reassure him, but the lump in my throat hurt more and more, and my eyes became moist in spite of my efforts.

I asked him to use the oxygen as little as possible so that we would have enough to last until the following morning. After explaining the situation to Philippe and Mama and telephoning to Touzé, who was anxious to come, I went back several times to make sure that Papa was resting, then I went to bed.

In my sleep I heard a shrill voice cry out in piercing tones, "BonPa is going to die! BonPa is going to die!" and I turned away, weeping and saying, "I know he will, but don't shout it so loud!" I would awake to hear his labored breathing in the next room, and I would become drowsy once more. And each time the voice began again, more harsh and shrill than before.

It is the only time in my life that I have had this experience, which is quite different from a dream, of a voice that speaks to you in your sleep. It is a terrible, frightening thing, especially when this voice repeats words that are so dreadfully sad to hear. . . .)

At Trousseau the X rays were judged to be very bad: two large nodules had appeared at the lower right where one week before there had been nothing. . . .

Birthe has always loved Papa with the affection of a real daughter, and that is what made so dear to him the atmosphere of our poor apartment where he came to lunch each week.

I went to look for Rémy at his house at two o'clock in the morning and luckily found him there. I explained the situation to him, and the three of us left in my station wagon.

I drove fast, very fast, with only one thought: "We must arrive on time. Papa must receive Extreme Unction on time. I promised him. This must be done. . . ."

When we arrived, his condition was stable, thanks to the oxygen that was now being administered continuously.

Around 4:30 the chaplain arrived, and the ceremony with Mama, Philippe, Rémy, and me kneeling at the foot of the bed was very simple and very touching.

When everyone had left and we were alone, Papa took me by the hand and said, "My son, now it is up to you to fight."

I nearly burst into sobs at these words, but I simply said that I would do all that it was humanly possible to do. God would decide.

I cannot recall exactly what we spoke about then while I held the inhaler for him. At one point he told me gently, between two breaths, "You know that that makes more than thirty years that we have lived together without ever a word between us?" . . .

The consultation that the three of us had in the large, cold, poorly lit sitting room was lugubrious. Touzé and Lafourcade appreciated the danger quite well but thought that if we could gain three days, the antibiotics perhaps would work.

After they had left I explained this to Papa. He raised his eyes to heaven and said, "Three days! I could never bear it for such a long time!" As Lafourcade was leaving I asked him to take Birthe back to be with our children. I did not dare to ask him to take Rémy back, too, without having a good reason. It was fortunate that I didn't, because that way the three sons did not leave their father again from that moment on. And that was good, because it was right that we, who had meant so much to him, should be near him at that time.

Today, one year later, I still thank God for having permitted me to be at my father's bedside during his final moments.

It was a dreadful trial, but this sorrow was pure and vivifying, and I am infinitely grateful to have experienced it.

Poor Papa, who was so anxious about my next voyage to America: "No one should travel so far from his elderly parents", he used to say. I know that my presence, though useless as a physician, was a great comfort to him, and for me, what he said to me, the way that he said it during those three days, is an immense treasure, summing up all the beautiful and simple things that he wanted to confide to me.

To be sure, I no longer remember everything, and the words that I transcribe here are only probably exact, perhaps not completely, but the calm, splendid, Christian way in which he waited for death while holding my hand is an unforgettable lesson that is wonderful, though dreadful.

That night I sat up beside him in the blue armchair, but

after a short time I came to give him some oxygen with the inhaler so that he could rest a little better.

At around eleven o'clock there were signs of heart failure: pulse slipping, low blood pressure, somnolence. Two cardiotonic injections, indicated by Touzé, performed one after the other at a half-hour interval, improved his condition, which remained stable during the course of the afternoon. I do not know how many times the good Doctor Touzé, quite upset, came to see him that day. That morning Philippe and I had obtained an ample supply of oxygen and had also brought from the hospital a tent, which Papa never wanted to use. He had a feeling of suffocating inside it. For nourishment Philippe suggested some Dextrosport, that sugar for athletes that Papa sometimes used, saying that to expend so much effort in breathing was actually like running a race. What a race, indeed, and toward what a finish line!

In the afternoon, Papa wrote on a piece of paper (which, alas, I have not been able to find) to ask for his maroon writing pad, and when Mama exclaimed that one mustn't speak of such things, he regained his breath for a moment, almost enough for his temper to flare up: "We have to look things in the face now."

At his request, I read to him the little text containing his last wishes, and he gave us his blessing. I treasure the document.

Kneeling at his bedside, I promised him then, with tears in my eyes, that if need be, everything would be just as he wanted. I gave him my word on it, and I kept it as well as I could. It was not easy, because Papa had asked that we not notify anyone, but we could not conceal his death from friends who telephoned or came by to ask news of him.

Nevertheless I asked our two best friends, Jean and Henri, not to come to the funeral at all. They understood this request, though it almost damaged our friendship with them.

At the end of the afternoon, when all three of us were at his side, he said to us, "How happy I am that you are here with me, all three of you! But I cannot say that to you all the time."

Toward evening, Touzé came by again.

His blood pressure, which had remained around 100, was now at 170, which for Papa was very high. It was a terrible shock to me. The two shots administered to him that morning came to mind: Was it too much? Had I erred all the previous times that I had taken his blood pressure? Or could this pinching of the arteries, this constriction of the veins, which now seemed so fine beneath his skin, be the final clash in this battle against asphyxiation?

In my room, to which I retreated for a moment, crushed, I begged God to let me not have made any mistake and not have unwittingly hastened the death of my father. It was simultaneously terrible and ridiculous, because I knew that there was nothing that could be done, that no medication could save him, and that those two shots had probably prolonged his life for a few hours. But the doubt was stronger; I feared that I had made a mistake.

Ever since that morning, regularly, inexorably, his respiration had been accelerating, and this panting became more and more superficial and weak. His pulse, too, was beating more rapidly. On the chart I was keeping, the numbers were higher with each passing hour. I got mixed up when counting the minutes on his watch, which was placed on his nightstand. I would count again, once, twice, three times, and it was still the same number, a little higher than an hour before.

Toward nine o'clock in the evening Touzé came by again. Papa made a sign and, raising both hands, said good-bye to us. The others did not understand. I had just let go of his hand so that he could make this gesture, and I almost shouted, "But don't you see that he is saying goodbye to you! Come and embrace him, before it is too late! Don't let him leave like that!" I don't know what else I said.

Each of us went to embrace him, and Papa then signaled to Touzé. This gesture overwhelmed the doctor, and he, too, went to bend over him; when he straightened up, he kissed me on the cheek while I sobbed.

Later there was an improvement, but I knew that it would not last. I sent Mama, Philippe, and Rémy to have dinner, promising to call them.

He held my hand and said to me, "You: you sure know how to love!" And when I replied that we all loved him, he corrected himself, "Yes, you are right, those are things that you don't say: you think them."

His heart weakened. A new shot of *solucamphre* revived him, and Philippe went downstairs in his turn. Suddenly, as I held his wrist, his pulse stopped. I hollered, "Mama, Philippe, Rémy!" They arrived—another *solucamphre*— and his heart started beating again. I recall having said then, "I will give you another shot to ease the pain, *mon Papa.*" And he said: "*Merci.*" . . . That was the last word that he spoke to me.

Then, during his agony, while Rémy held the oxygen mask, we knelt down and started to recite the Rosary. Visibly, even though he could not speak, Papa was saying it with us.

He was sweating, felt cold and then hot; I wiped his forehead and his chest, which were covered with the sweat of his agony. Philippe asked him in a faint voice, "Do you know where you are going?" He nodded yes, as though annoyed by a question about something obvious.

During the long hours when we had been alone, I said to him repeatedly, "My Papa, my dear old Papa, how it grieves me to see you suffering like this!" And he would answer me with a fluttering of his eyelids.

Late in the night, on Saturday, January 11, 1958, at about 12:30 A.M., his breathing became even more superficial, and the pulse faded again. My syringe was ready, but I did not want to let go of his hand. I wanted to keep his dear hand in mine until the final moment.

He wanted to make the Sign of the Cross, but the oxygen mask got in his way, so I told Rémy to remove it for a moment.

After consulting Philippe with a glance, I gave him a final injection, which I knew was useless, but his heart had rallied two times already, and the second time I had said to him, "You have a strong heart, Papa!" And in saying that I was not thinking so much of the organ as of his fatherly heart.

Every time he inhaled the respiration was more shallow. Finally he caught a bit of breath, and I heard him murmur something—I understood "Jesus"—and I remembered the grace of the plenary indulgence *in articulo mortis* that the Pope had granted to our family.

Then, with his hand in mine, as he had imagined for Finkerman, and as he had done for his father, he gave up his spirit.

Just before that final moment, I leaned over to his ear —I am almost sure that he could still hear me—and I said to him, "Adieu, *mon Papa*, and thank you."

That is all that I could think of to say to my dear father, whom I loved with all my soul and whom I could not save. I thanked him for our whole life, for his goodness, for our friendship, and for his kindness even to these final moments. He had said to me in the afternoon, "If it weren't for the hope of seeing each other again, this would be terrible."

I have hope of seeing him again, and, if God grants it, we will meet again one day.

What happened afterward I only recall vaguely.

I know that Mama gave me the wedding ring that they

removed from his hand, and I keep it with me always. I know also that Mama told me to close his eyes, that after what I had done for him, it was for me to close his eyelids for ever.

My dear Papa, whom I loved with all my heart, with whom I shared love and understanding during many long years, who was for me an advisor and a friend, my dear Papa is no more.

Often the thought occurs to me, "Well, of course, I will ask Papa about that"—and then I correct myself and, rephrasing the question, I ask myself, "What would Papa have said about that? What would he have done?" and it is the memory of his whole life that then offers me the advice that his dear voice can no longer give me.

That was the end of our wonderful walks on the hill with the children, the end of the little "odds and ends of business" at his office or at the bank, the end of our Thursday dinners, the end of all that happiness that we both enjoyed fully and that was one of the most beautiful gifts that Providence has given me.

Now his body is resting in the little tomb in the cemetery of Montparnasse where my grandparents are buried, which both of us used to visit each year.

Now I go back there alone, often, to be near him for a little while yet, to pray to God for him and so to give witness that he was a very good father, and to ask God to take him to himself.

Message of Pope John Paul II

*I am the resurrection and the life; he
who believes in me, though he die,
yet shall he live. (Jn 11:25)*

These words of Christ come to mind when we find ourselves faced with the death of Professor Jérôme Lejeune.
If the Father who is in heaven called him from this earth
on the very day of Christ's Resurrection, it is difficult
not to see in this coincidence a sign. The Resurrection
of Christ stands as a great testimonial to the fact that life
is stronger than death. Enlightened by these words of the
Lord, we see the death of every human person as a par-
ticipation in the death of Christ and in his Resurrection,
especially when a death occurs on the very day of the
Resurrection. Such a death gives an even stronger testimony to the life to which man is called in Jesus Christ.
Throughout the life of our brother Jérôme, this call was
a guiding force. In his capacity as a learned biologist, he
was passionately interested in life. In his field he was one
of the greatest authorities in the world. Various organiza-
tions invited him to give lectures and consulted him for
his advice. He was respected even by those who did not
share his deepest convictions.

We wish today to thank the Creator, "of whom all paternity in heaven and earth is named" [Eph 3:15 (Douay-Rheims)], for the particular charism of the deceased. One must speak here of a charism, because Professor Lejeune was always able to employ his profound knowledge of life and of its secrets for the true good of man and of humanity, and only for that purpose. He became one of the ardent defenders of life, especially of the life of preborn children, which, in our contemporary civilization, is often endangered to such an extent that one could think the danger to be by design. Today, this danger extends equally to elderly and sick persons. Human tribunals and democratically elected parliaments usurp the right to determine who has the right to live and, conversely, who could find that this right has been denied him through no fault of his own. In different ways, our century has experimented with such an attitude, above all during the Second World War, yet also after the end of the war. Professor Jérôme Lejeune assumed the full responsibility that was his as a scientist, and he was ready to become a "sign of contradiction", regardless of the pressures exerted by a permissive society or of the ostracism that he underwent.

We are faced today with the death of a great Christian of the twentieth century, of a man for whom the defense of life became an apostolate. It is clear that, in the present world situation, this form of lay apostolate is particularly necessary. We want to thank God today—him who is the Author of life—for everything that Professor Lejeune has been for us, for everything that he did to defend and to promote the dignity of human life. In particular, I would like to thank him for having taken the initiative in the creation of the Pontifical Academy pro Vita. A long-time member of the Pontifical Academy of Sciences, Professor Lejeune made all the necessary preparations for this new foundation, and he became its first president. We are sure that henceforth he will pray to the Divine Wisdom for this institution, which is so important and which in large measure owes him its existence.

Christ said, "I am the resurrection and the life. He who believes in me, though he die, yet shall he live." We believe that these words have been accomplished in the life and in the death of our brother Jérôme. May the truth about life be also a source of spiritual strength for the family of the deceased, for the Church in France, and for all of us, to whom Professor Lejeune has left the truly brilliant witness of his life as a man and as a Christian.

In prayer, I unite myself with all those who participate in the funeral, and I impart to all, through the mediation of the Cardinal-Archbishop of Paris, my apostolic blessing.

The Vatican, April 4, 1994